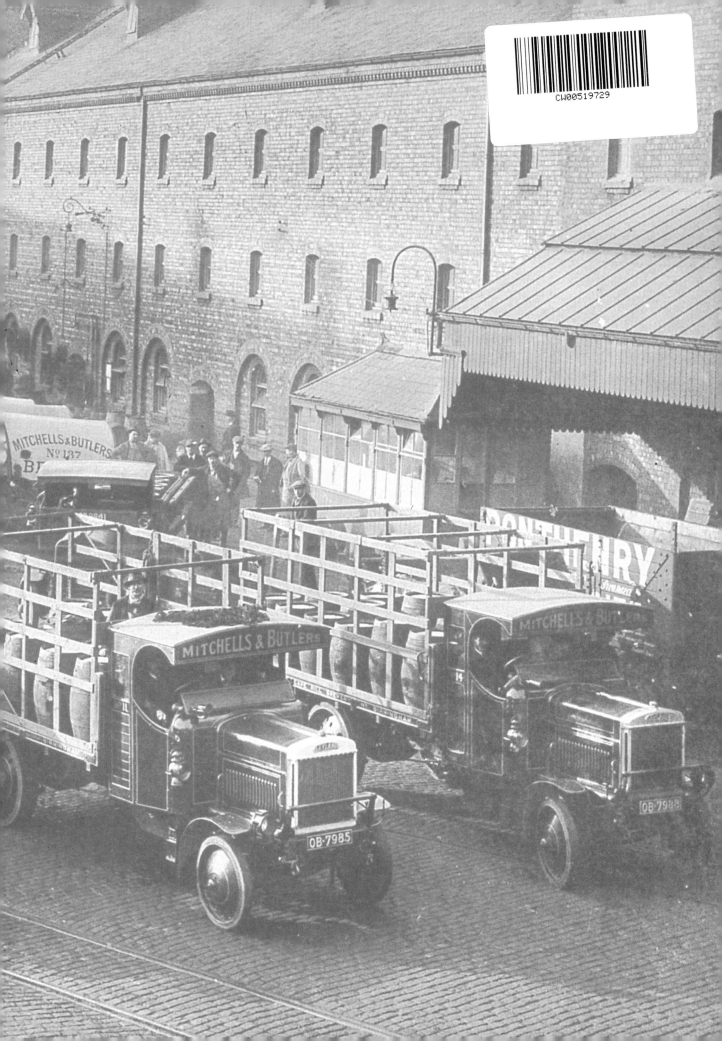

Leyland LORRIES

A Celebration

ROUNDOAK PUBLISHING, NYNEHEAD, WELLINGTON, SOMERSET

First published in 1996 by Roundoak Publishing
Nynehead, Wellington, Somerset, England TA21 0BX

ISBN 1 871565 25 1

Typesetting: Character Graphics, Taunton

Printed in England by
The Amadeus Press, Huddersfield

Leyland Lorries has been published in association with the
*British Commercial Vehicle
Museum Trust Archives*

The majority of the photographs in this
publication are from the Archives of the
British Commercial Vehicle Museum Trust who
retain full copyright of the photographic
material supplied.

Additional photographic material has
been supplied by the Editor,
Arthur Ingram

Above: This 1951 view of Shotton steelworks at Connah's Quay near Chester, shows both prewar and postwar examples of the Leyland 'Octopus' in service with John Summers & Sons Ltd. The 1937 TEW8T model on the right was one of 14 operated in prewar days, while the 1950 example of a 22.O/1 was an early delivery of the total of 84 taken into service since 1946.

Front Cover and opposite: A Leyland 'Octopus' of Naylors Transport (Leyland) Ltd captured heading south over, what was the then, the last Thames crossing in January 1958. A popular venue for photographers – professional and enthusiasts alike – Tower bridge, which celebrated its own centenary in 1994 provided a convenient crossing point over the Thames for heavy lorries until the imposition of weight restrictions in 1979.

Leyland
LORRIES

Editor

Arthur Ingram

CONTENTS

A mid-fifties view of the A6 trunk road where it passes through Penrith, a view remembered by many old long haul drivers. The Leyland 'Steer' of Cargett & Wilson of Penruddock is about to pass through 'the Narrows' going north, which thankfully still remains largely unaltered today.

INTRODUCTION

The photographic collections within the Archives of the British Commercial Vehicle Museum Trust, constitute a major industrial and socio-historic resource charting the development of the British commercial vehicle industry.

One of the purposes of this publication is to reveal examples of the rare and unique contents, of a photographic archive of national importance. The majority of the photographs included in this book are from just one of those collections, namely that of Leyland Motors Ltd.

Changes in managements within the Leyland Group of companies, has resulted in its photographic collections being neglected. In order to ensure the survival of this record of our industrial past, steps must be taken to preserve the collection prior to its being fully exploited. Therefore, income generated from the sale of this book (which, it is hoped, is only the first in a series of future publications), is being used for the promotion of the long-term preservation of these irreplaceable photographic images.

It is hoped that the book has a universal appeal, and for this reason the technical descriptions, typical of specialist books, have been avoided. Furthermore, photographs of vehicles in their natural working environments have been chosen wherever possible, because of the added appeal of period scenes.

Roslyn M. Thistlewood CertEd DAA
March 1996

Below: The old-established Manchester brewers Boddingtons, have achieved considerable attention recently because of heavy advertising under Whitbread control. Way back in the Autumn of 1935 they took delivery of this TSA4 'Badger', and it was photographed in Chorley, passing the house of a competitor: Chesters. Interestingly, the upper storeys of the Royal Oak House now house the Archive of the British Commercial Vehicle Museum Trust.

FOREWORD

Until my recent retirement in 1995, I had been employed in road transport and in the transport manufacturing industry, for over 30 years. For the last 20 years I have been involved in the collection and preservation of historical commercial transport records, mainly from the companies taken over by the Leyland Group. These records are now in the custody of the British Commercial Vehicle Museum Trust Archives.

Many people over the years have expressed a desire to be given both detailed information about, and access to, the Trust's collection. However, they do not realise the amount of work that has to be carried out (involving sorting, conserving, listing, boxing and labelling) on such a vast collection, before it can be made available for research. Even though much of the collection is now accessible, there is still a considerable amount of work to be undertaken for many years to come.

One of my ambitions, to make publicly available some of the photographs from the unique photographic collection of Leyland Motors Ltd. is about to be realised in 1996, with the publication of this book. It is appropriate that we present this insight into a small part of the collection in the year celebrating 100 years of commercial motor manufacture in the town of Leyland.

It is my earnest hope that this book marks just the start of a series of publications, which reveal the magnificent contents of the Trust's Archives. Other photographic collections planned for release in future publications are those of AEC, Scammell and Thornycroft. However, there is much to be achieved in their conservation and preservation, before they are made readily available.

Gordon Baron

EDITOR'S NOTES

I first visited Leyland over 30 years ago, and was glad to find that the majority of the photographic records of the company still existed. As I viewed the images of lorries long forgotten, from the GH4, SQ2 and RAF, to Badger, Rhino and Cub, I became hooked on Leylands.

So now, after countless new models and types, and almost as many years, it is gratifying to be going through some of these photographs and contributing to this volume of Leyland Lorries.

Regular readers will know of my penchant for 'street photographs' and Gordon Baron has searched the files for as many action photographs as possible, ably assisted by Dave Lewis who was responsible for reproducing many of the images. We all know how the vehicle manufacturers liked to pose their shiny new products in rather bland locations, and Leylands were no exception, for the solitude of Worden Park was just so convenient.

We called upon the expertise of Neil Steele to sort the chapter on fire engines, as an ex fireman and an ardent Leyland fan, he had to be reined in a little on his choice. Anyway, he says, Leylands were responsible for more than half the British fire engines in prewar days – so there!

The main thrust of this book is to provide a tiny glimpse of what can be found in the BCVM Trust Archive with regard to lorry production, although this has been stretched to take in fire engines, plus a brief look at wartime vehicles. In no way have we set out to provide a detailed history, full of technical facts and figures. The other specialised fields of the Leyland cars of the 1920s, the petrol trams, the railcars and the Trojan, have all been purposely excluded and deserve detailed investigation in their own right.

Almost every photograph included, comes from the Trust Archive at Chorley, with just a handful being added from my own collection for balance.

In addition to those named above, I would like to take this opportunity of recording my grateful thanks to others who came up with various information, dates and details when asked. These include Malcolm Wilford, George Wrey, Imperial War Museum, The Shuttleworth Collection.

I must also pay tribute to the publishers of Commercial Motor, Motor Traction and Motor Transport, for over the years they have provided us researchers with enormous amounts of information to enlighten our way.

Arthur Ingram MCIT

THE EARLY YEARS

With ancestry going back to 1884 when James Sumner built his first steam vehicle, the Lancashire Steam Motor Company was formed in 1896, with three members of the Spurrier family joining him at the Herbert Street works in Leyland.

The first actual sale of a steam vehicle was recorded in 1897, and a few others followed in the years round the turn of the century. The first export was recorded, an improved model introduced, the works extended and additional workers employed.

The first petrol engine vehicle was built in 1904, a better design followed and during 1905 a total of 16 chassis was built. In 1907 the steam vehicle competitor, Coulthards from nearby Preston was bought out, and the new combine renamed Leyland Motors Limited.

The next two chassis designs really gave substance to the production of petrol vehicles, the 35hp X-type and the larger U-type with a 50hp engine. With these two petrol models and the existing steam wagons, Leylands were able to offer a range which it hoped would attract many orders.

Further extensions to the works ensued, the first fire engine built in 1909, and in the following year a shooting van for King George V was to lead to the granting of the first Royal Warrant to Leyland Motors Ltd.

In 1912 the War Office set up vehicle trials to determine the suitability of petrol vehicles for their Subsidy scheme, In future owners of these approved subsidy type vehicles would receive a purchase premium of £50, and a subsidy of £20 for each vehicle for a period of three years. In order to qualify, vehicle owners had to maintain them to the satisfaction of War Office Inspectors. In the event of mobilisation of

Below: A nice period photograph showing one of the early steam wagons actually in steam in about 1904. This was the period when many contractors were turning to mechanical transport for the first time, and quite a number of the early Leyland photographs show steam wagons with fleet number 1 on the apron. This 6 ton tipper, working for Andrew Knowles of Pendlebury, near Manchester, appears to be loaded to capacity.

the Army Reserve, the War Office was entitled to purchase the vehicle at its current value plus 25%, and the owner had to deliver it within 24 hours.

Two Leyland vehicles were entered for the War Office Trials, one in each of two classes: 30cwt and 3ton. In the event Certificates were awarded to both the Leyland types entered. In the year between April 1912 and April 1913, the War Office placed orders for 88 vehicles. Later orders called for the 3ton type and the 30cwt model played a minor role.

Above: It is almost unbelievable that ammunition would be carried in a steam vehicle, but we must assume that suitable safety precautions were taken. This steam wagon was photographed on the railway bridge at Leyland, before being delivered to Eley Brothers at Edmonton, a company which still produces shotgun cartridges. The rear extension to the body is highly unusual, and was probably specified as protection when cartridges are being unloaded in the rain – you must keep your powder dry! The slight imperfection of the photograph indicates the problems of decomposition of the emulsion on some of the old glass plate negatives.

Right: Posed adjacent to Leyland railway station and the Railway Hotel, this early petrol engined vehicle shows the simple layout of the 1905 product. The chassis frame is of rolled steel section, with the road spring brackets on the outside. The front wheels have rubber tyres, but the rears are still of the old steel rimmed type so typical of early steam vehicle design. The 40hp engine is protected on the underside by a mudshield, which also contained any oil leaks, while the hinged bonnet carries a brass plate proclaiming that the design by the Lancashire Steam Motor Company is patented. The fluted radiator top tank is reminiscent of some early American trucks, while the absence of any protection for the driver was normal for these early machines.

Inset right: The 3 ton X-type was the first model produced in any numbers. Introduced in 1907, it was fitted with a six litre, four cylinder engined rated at 35hp and produced by Crossley Motors of Manchester. This 1909 vehicle was photographed on the railway bridge at Leyland, and the very deep radiator with the starting handle running through the middle, is immediately obvious. By this date most vehicles had some form of cover for the driver, which might amount to just a canvas cover. The vehicles now had rubber tyres all round, including twins on the rear axle.

Above: Photographed later in its life, this pre Great War model is in immaculate condition as seen leaving the Broad Green, Liverpool works of the English Margarine Works in 1929. The driver has even gone so far as wrapping his acetylene headlamp and oil sidelamps in cloth to keep them clean! By this time glass windscreens with an opening section were becoming the norm on commercial vehicles. Note that a model aeroplane mascot is carried on the radiator filler cap, and the Leyland script name has been added, just below the large oval Lancashire Steam Wagon Company badge.

Above, left & right: In 1912 Leylands submitted two models for the War Office Subsidy Trials: the type A for a 3 ton load, and the smaller type B rated as a 1½ ton load carrier. These two illustrations come from a contemporary sales brochure, because of the loss of the original negatives. A comment by the trade press of the day mentions the annoyance registered by certain manufacturers about the fact that the Leyland designs had been readily accepted by the War Office committee. They then formed the basis for the W.O. specification for all vehicles, and this was resented.

Below: Carter, Paterson & Co., the parcels carriers, began as far back as 1887, trying out some of the early motor vehicles around the turn of the century. They tried out six Leylands in 1909, liked them and went on to buy numbers of X-, S- and G- types in the years which followed. The vehicles completed high mileages during their lifetime, but were well maintained; old number 137 has survived and is now on show in the BCVM at Leyland.

Above: Almost from its inception, the Leyland works was fortunate in having many famous names grace the order books. In the world of furnishings, both Waring and Maples were customers, and one of the latter's vehicles was used to illustrate the 5 ton model. This particular design was known as the 'well van', because it was constructed with a well in the van floor at the rear, between the specially-wide chassis rails. This feature was advantageous for the loading of heavy items of furniture from the ground. A price of £960 was quoted for this vehicle, in the days before the Great War.

Above: At the lighter end of the range were the 15 cwt and 1 ton models, one of the latter being shown here, in the livery of Chiesman Brothers, one of the large departmental furnishing stores, which still exists today in Lewisham. One standard chassis formed the basis for both 15 cwt and 1 ton models, with 14hp and 18hp engines being used as required, these being of Aster manufacture. A small factory-built ambulance was also advertised, using the 1 ton chassis.

Left: Looking decidedly ungainly, with the driver positioned way above the engine, the ancestry of the 5 ton 'overtype' Leyland, is plain to see. As the precursor of the modern forward control truck, the great selling point of the design was the greater body length available by moving the driver to this elevated position, above the engine compartment. It did not find universal acceptance however, and the more reasonable layout of 'driver-beside-engine' which Leyland called the 'side type' followed later, and marked the true move away from the conventional bonnetted layout.

Below: The London furnishing store, Waring's, caused something of a furore early in 1913 after they purchased eight Leylands which were loosely described as the 'aviation type', only to discover that the War Office inspector refused to grant them approval as subsidy models. Under protest he relented, much to the dismay of other manufacturers whose designs had not met with his approval. Here, one of that batch is about to set off for Avonmouth docks for work with the Army Service Corps. Note the long chain carried just below the Waring's name, a necessary fitment for vehicles engaged on subsidy service for emergency towing.

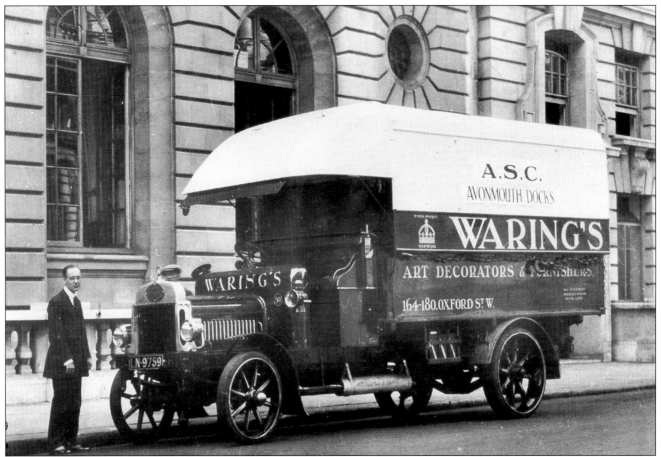

THE GREAT WAR LEYLANDS

As the originators of both 'A' and 'B' type War Office Subsidy models, Leylands were well positioned for the production of vehicles at the outbreak of the Great War.

The company was re-established as Leyland Motors (1914) Ltd, with Henry Spurrier the first as Chairman, with his son, Henry the second, as Managing Director.

Up to the outbreak of war in 1914, Leyland had built 2092 petrol vehicles, and these had established an enviable reputation for quality and longevity.

Within 24 hours of the Declaration of War on August 4, numerous fleets of registered subsidy vehicles were recalled to the factory from their civilian owners, where they were prepared for their new role as war machines. Within another 24 hours large numbers of vehicles were on their way to the embarkation depot at Avonmouth.

So, after just a few hours of war, Leyland Motors had begun its considerable contribution to the national war effort. The first wartime military vehicle was delivered on 14 August 1914.

In 1915 the War Office decided that the entire Leyland production should be allocated to the Royal Flying Corps, and a Training Centre for equipment officers and air mechanics of the RFC was instituted.

Within a short while, Leyland chassis were being turned out for service as a variety of vehicle types for the RFC, based on the 3 ton aviation chassis as it was called, although in later years it was referred to as the 'RAF type'. Records show that almost 6000 of this 'A' type subsidy chassis were produced during the war period.

A selection of the vehicle types supplied to the RFC (later RAF) is shown in the photographs which follow, but in total they included mobile workshops, bomb carriers, AA gun platforms, portable searchlight units, petrol tankers, aircraft transporters, dentist' vans, kite balloon winches, radio & communications vehicles and general tenders.

Above: Posed alongside this Bristol F2a biplane in unfinished form, is one of the many RAF types allocated to the transport of aircraft in sling vans. These vans or containers were rather longer than the load space of the vehicle, but the load was not unduly heavy, consisting of aircraft in 'knock-down' form, for assembly at service locations.

Above: Most of the Leylands employed by the British air forces in the Great War, were of the general service or tender type, but there were many other varieties, such as this tanker for aircraft refuelling. The standard 3 ton chassis/cab carried a welded steel tank which was secured by adjustable steel straps. A small tool box was positioned behind the cab, and a hand pump and fire extinguisher were housed in the covering for the outlet cock at the rear.

Below: This photograph shows the signals van, with a portable generating set to power the wireless equipment. One wonders how the operator could hear messages, with the generator engine positioned just at his back. The vehicle carries a designation for Mesopotamia on its side.

Below: One of the everyday general service tenders, modified by the addition of an Aster-powered winch and winding drum, for the control of aerial kites or balloons.

Above: The rough wooden body on this RAF-type, has been lengthened to accommodate fuselage and wing sections of damaged aircraft. Here a Bristol SE5 in parts is being taken back to an RAF depot for salvage or rebuilding.

Above: With the cape cart hood extended, this mobile workshop is providing facilities for the fitters working on the DH9s at an RAF station 'somewhere in England'. It will be noted that the equipment includes a small petrol engine, driving a dc generator to power the lathe and pillar drill. All three sides of the body are hinged outward to increase the work area, and canvas curtains can be rolled down in the event of inclement weather.

Above: A RAF-type tender is pressed into service as a recovery vehicle in order to right this plane, which has suffered a mishap on landing. With nine men to help, the driver waits for the OK to put the plane back on its wheels, with the rope attached to the tail skid of the plane, and the other end tied to a length of timber tied across the front of the tender's body.

Above: A fitting testimony to the RAF type is provided by this photograph of a service vehicle based at Farnborough, with its capacity load, The outsize packing case is labelled for the Aircraft depot, Karachi, India, and judging by its weight, contains more than just one aeroplane. Upon closer inspection, the photograph reveals that the vehicle has no proper body, the container being roped to a number of planks laid across the chassis.

THE 1920s: REBUILDING & REJUVENATION

With the end of the Great War, the country was in a far different state than in the years preceding the conflict.

Although there had been terrible loss of life in Europe, there were still many hundreds of thousands of men returning home with great expectations for the future. Transport had played an important part at the battlefront as well as in a support role. Many soldiers had gained a taste of mechanical transport, and were anxious to translate that into a living for themselves.

Not all wanted to be just lorry drivers, many had the urge to try their hand as a haulier, and there were plenty of lorries available. Thousands of military vehicles were being stockpiled in fields and dumps, ready to be auctioned off to the highest bidder. Many of these vehicles were beyond redemption, but there were others that could soon be put back into use, so a period of rebuilding began.

Leyland were apprehensive about the situation, they could see that the market for new vehicles was being swamped by a mass of cheap but perhaps unreliable equipment. In addition to this, there were plenty of Leylands among those on the dumps, and the Leyland reputation was at stake.

So the decision was made to acquire as many ex military Leylands as possible, and rebuild them to a high standard, before offering them for sale.

Ham works, near Kingston-on-Thames was secured from the Sopwith Aviation Company, and the task of refurbishing began. The vehicles were sorted, and those beyond economic recovery, dismantled. Those considered worthy of a rebuild were completely stripped, and all the various parts and assemblies remachined as necessary, to the Leyland standard.

The majority of the refurbished RAF types were sold as 4 ton lorries, with just a few finding use as buses. Rebuilding carried on until 1926, by which time some 3000 had passed through the Ham works.

The cost of this vast rebuilding exercise was considerable, and there was plenty of competition from the hordes

Above: A scene inside the Ham works at Kingston-on-Thames, with work proceeding on the assembly of a rebuilt RAF type chassis early in 1923. With a white-coated foreman watching the photographer, the two fitters appear to be securing the spring hanger brackets. The remainder of the workshop seems to be devoid of any activity.

of other ex government chassis being available at far less than the £590 originally asked for the RAF type. The company went through a very lean time, and many people thought it would all lead to disaster, but it came through.

The 1920s was not all RAF types. Other models continued to be offered, from the short-lived 30 cwt, to the 7 ton SQ model, plus the extra capacity six-wheelers and the articulated Leyland Carrimore.

Things really began to improve toward the end of the decade, when J.G. Rackham's designs for the T-engine range gradually came into production. The bus market was tackled first, but soon after came the lorry designs and the market was offered chassis which better met current demands.

The new ranges of both bus and lorry chassis were given animal names, which came to mind far easier than a series of chassis letters. So the transport world was beset with the

'Badger' and the 'Beaver' for a start, with the promise of others to follow on later.

The closing years of the 1920s set the mould for the next decade so far as Leyland was concerned, for there was to follow a period of expansion and refinement to the marque. But it would also be a period of tremendous changes, effecting not only the vehicle manufacturers, but the whole concept of road transport in Britain.

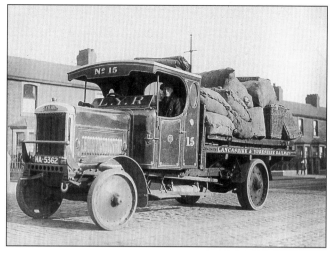

Right: Not surprisingly, the Lancashire & Yorkshire Railway employed a considerable number of Leylands for their collection and delivery fleet. Another instance of their progressive ideas in transport was the use of 'Lancashire flats' for the speedier handling of loads. These 'flats' were similar to a stout platform lorry body with four hooks or eyes, so that they could be craned on and off for loading or trans-shipment. Leyland Motors produced a series of these 'flats' in four different sizes to suit the C, G, P and Q models in their range of the 1920s.

Left: One of the many reconditioned 'RAF' type 3-tonners, designated as 4-tonners for the civilian market, and finished in publicity livery. Leylands tried hard to maintain their reputation for quality and reliability, by rebuilding the vehicles to an 'as new' standard, a practice which almost bankrupted the company.

Below: Everyone in this photograph was keeping perfectly still for the long exposure required to capture this RAF type, loading inside a mill. The driver and his mate look almost military in their uniform caps, but this was the style adopted by biscuit makers Peek, Frean and Co. for their transport personnel in the 1920s. The vehicle retains a utilitarian appearance with its canvas hood for the crew, and has no provision for any headlamps.

Above: An evocative early morning view of the despatch yard at Mitchells & Butlers' Cape Hill brewery in Birmingham in 1920. Pride of place is given to the recently-acquired GH4 models, some of which have post-and-chain bodies for loads of heavy casks, while others are fitted with high slatted sides for capacity loads of empties. The number of horse-drawn vans was still considerable at this time, as was the use of railway wagons for the supply of coal for the brewhouse boilers.

Right: It is gratifying to find that some of the very old photographs in the Leyland archive show vehicles actually in use, as opposed to the carefully arranged publicity shots so often encountered. Here we can witness the considerable activity at Clitheroe railway goods yard, with local dairy farmers bringing their few churns of milk to meet the GH2 lorry of Whalley District Farmers co-operative, for transfer to the creamery, in 1920.

Above: The 'Q' model was rated as a 6 ton machine and boasted the largest engine of the period: the 40/48hp. In this 1921 photograph an example in service with the County Palatine Motor Engineering and Transport Co. of Blackburn, is shown alongside a steam freighter at Preston docks. The two men at work on the lorry appear to be complete with jackets and neckties, while the pair on the railway wagon are down to their waistcoats.

Right: The quality of the old glass plate negatives in the Leyland archives is clearly evident in this 1920 photograph of a GH4 model for the Co-operative Wholesale Society at Manchester. The vehicle is finished in a plain but immaculate livery, with highly burnished aluminium radiator and acetylene headlamps, the latter being fed with the gas generated in the unit just behind the front mudguard. The driver has a padded seat cushion and backrest, the windscreen is set in a polished hardwood surround, but access to the cab is from the nearside only because of the gear and handbrake levers being close to the offside.

Left: The Preston-based haulage firm of H. Viney & Co were an old established contractor in the area, beginning in business in the last century, and formed as a limited company in 1906. The Preston office was in Strand Road, and they had branches at Manchester and Liverpool. The company was loyal to Leylands for the majority of its fleet right up to nationalisation in 1949. This 1921 photograph shows a 'Q' type sixtonner, complete with drawbar trailer, and the loosely sheeted load appears to be cotton waste. Of interest is the fact that the lorry body is carried on shaped 'stools' in place of the more usual longitudinal body runners which raise the body floor clear of the rear mudguards.

Below: This view of a 1924 model A1 two tonner, shows how the current arrangement of scuttle and windscreen was adapted to the design for integral van bodywork. Barker & Dobson proudly proclaim their Viking chocolates on the nearside, and their equally famous Everton toffee was probably displayed on the offside, and interestingly the company name still exists today.

Left: The name of Irwins was well known in the north-west, for the company ran a large grocery chain business on Merseyside and in North Wales for many years up to the 1960s when they sold out to Tesco Stores. Here we see one of their PH2 five ton Leylands, posed prior to delivery in 1924. Note that the solid rubber tyres on the rear axle are not pairs or twin tyres as is more usual, but no less than four sections arranged across each wide base steel wheel.

Below: At the lighter end of the petrol engined range was the A1 type rated as a two ton load carrier, although from this photograph we can see that the vehicle weighed almost 3¼ tons unladen. Specification for this type lists a four cylinder engine of 30-32hp driving through a four speed gearbox, to a worm final drive rear axle. The cab is the standard timber construction for the period, with a door on the nearside only, and the platform type body is fitted with detachable stanchions and chains to contain the load. Difficult to distinguish in this 1923 photograph is the Royal Cypher mounted atop the radiator filler cap, and the Instructions to Drivers notice displayed inside the cab.

Above: The articulated sixwheeler did not feature to any great extent in transport fleets until 1920, with the advent of the Scammell. Following the success of this integrated design, several vehicle manufacturers sought to enter the market with matched semi-trailers. Leyland's answer to the competition was an arrangement with trailer builders Carrimore using a short wheelbase version of the current four ton chassis. This 1924 version of the design is pictured here with a considerable load of cast steel compressed gas cylinders – try counting them! Later to be known as the Leyland-Carrimore Lynx, it featured pneumatic tyres all round, but sales were disappointing, operators preferring the lorry-and-trailer outfits. One disadvantage of these early articulated designs was the absence of any quick and easy method of detaching the trailer.

Left: The 'C' type was rated at three tons capacity and was generally similar to the 'A' type two tonner using the 30-32hp engine, although a longer wheelbase was offered and it weighed a little heavier than its junior stablemate. The substantial fixed sided body has a sturdy front gantry and intermediate trestle to support long loads encountered in the building industry. This 1923 photograph from the BCVM Archive, shows some signs of deterioration in the emulsion on the glass plate, and highlights the problems faced in the monumental task of preserving the ageing collection of such valuable historic material.

Above: Leylands never seemed to be very successful at the lighter end of the payload scale in vehicles – they were probably too well engineered, and hence too expensive. The 'Z' type was the company's offering as a one and a half ton load carrier in the mid 1920's and its relationship to the heavier models is plain to see in this example for work with a Halifax based brewer. The little engine produced 20hp, but it did feature an overhead camshaft layout, while another advantage was the fitting of pneumatic tyres as standard.

Right: One of the more impressive vehicles of the mid 1920s production was the 'Q' type six tonner at the heavier end of the payload scale, and its successor, the semi-forward control version, the SQ2 6/7 tonner. This 1926 example was operated by the South Birmingham Transport Co. of Selly Oak, which was established in June 1920 under the direction of George Samworth and his son. Although this vehicle has achieved the distinction of electric lighting – yes, those Lucas sidelamps are adapted – it still retains features of a decade earlier. Indeed the SQ model was still in the catalogue after the introduction of the T-range models in 1929.

Above: Ever mindful of the increased earnings through greater carrying capacity, many transport operators looked to the rigid sixwheeler as their salvation. There were several add-on axle options available, but naturally the vehicle builders were not keen to see their original chassis suffering from gross overloading because of claims by axle salesmen. Leyland chose the path of lengthening the chassis frame of the SQ2, and adding a third axle <u>ahead</u> of the normal rear axle. In order to counteract the tyre scrub normally associated with the rigid sixwheeler, the additional middle axle was made to steer. Mrs. M. Smith & Sons of Kidsgrove was one operator who tried the new 'Longframe' model, this 1929 photograph showing one of the 10 ton vans of that fleet.

Above: Established in 1896, Lloyd's of Manchester started using motor vehicles in 1914, and were one of the first operators of the new TA1 'Badger' model in 1929. From this photograph we can see clearly the earlier parentage in the shape of the radiator, bonnet and driver's cab, but a lower chassis and steel disc wheels with pneumatic tyres, help give a more modern appearance. It would appear that this driver was given notice of the arrival of the photographer, and has been busy with the white paint!

Left: Testimony to the longevity of Leylands, is provided by some of the vehicles built in the early 1920s. Many lasted well into the postwar years, as the veteran shown here in service with Maples the furnishers. Grudgingly accepting progress in the form of a windscreen and pneumatic tyres fitted in the 1930s, this individual was still clinging to its acetylene headlamp and oil sidelamps when photographed outside Olympia in May 1951.

THE 1930s: EXPANDING THE RANGE

The 1930s was to witness a revolution in the road transport industry. Commercial vehicles, designed for the efficient and economic movement of all types of traffic, were being seen more and more as the prime mover of the nation's economy. Much to the annoyance of the railway companies.

Vehicle design leapt ahead, helped by improvements in metallurgy, materials, construction and specification, but legislation began to make itself felt in both the design and operation of the commercial vehicle.

Leylands expanded their market share with the introduction of new models, which were then constantly reviewed and revised in order to keep them abreast of the competition from makers such as AEC, Albion, Atkinson, ERF and Foden.

The Ham works took up the production of a completely new lightweight range – the 'Cub', and later the 'Lynx'. While in Leyland the menagerie was extended with 'Hippo', 'Bison', 'Bull', 'Buffalo' and the first oil engine chassis, the 'Rhino'. Shortly the 'Octopus' and 'Steer' names would join the family.

The whole range of vehicles exemplified the position of Leyland in the market place. There was a model available for almost every transport task, save that of the very small or the most specialised. The handsome, polished radiator shell; the comfortable cab; the high specification; all went to produce a vehicle of exceptional value and service.

During the decade, the military side of production was catered for by the introduction of the 'Terrier' and 'Retriever' six-wheelers, but Britain was not in the business of building a huge arsenal, so orders were small. Later, the threat of war was to make itself manifest by Home Office orders for power and equipment to tackle the expected air raids on the home front, but the general switch to war production was yet to come.

Above: Into the 1930s the TA1 'Badger' was given a new style of cab with full height doors with sliding windows, which helped to shed the 1920s look, this example for the Horwich Industrial Co-operative Society even boasting a pillarless windscreen. The simple style of bodywork with hoops and rails plus a throw-over sheet so typical of Lancashire, is utilised to cover the motley load of crates, casks and baskets.

Left: The TC3 model 'Beaver' was listed as the 'long goods' in the Leyland data sheet, with a wheelbase of 14ft 6in. This example for Joseph Lucas with its integral bodywork, was certainly an elegant vehicle, from its gently curving roofline, down to the continuous running board. Delivered in March 1930, it seems incredible that the payload was a mere 3½ tons.

Right: During the 1930s Worden Park, Leyland, was the location adopted for photography of new vehicles, and on this occasion in July 1936 a TSA4 'Badger' is the subject. By now the range was decidedly modern, and they have even been described as 'handsome' by some dedicated enthusiasts. The deep, polished aluminium radiator shell, with its Leyland badge and model name proudly displayed, was certainly impressive, particularly perceived from close range. This design of well-supported dropside body is typical of the period when many haulage contractors were called upon to carry almost any commodity at short notice. With the side down the load was at shoulder height, and with the side up a lot of items could be carried without recourse to sheets and ropes.

Left: Both the TA 'Badger' and TC 'Beaver' were popular with brewers, because the cabs of these bonnetted models could accommodate the three-man crews often used for heavy cask work. Huddersfield-based brewers Bentley & Shaw purchased a batch of six TA3 'Badger' models in 1933, and one is shown here delivering to a public house in Northumberland Street, Huddersfield in May 1935. The driver can be seen sorting the load, while the two trouncers prepare the casks for lowering into the house cellar.

Above: The TA series 'Badger' was rated as a 2½ ton payload machine when first introduced in 1928, but it was soon uprated to become a 3 tonner, a 4 tonner, and finally 6 tonner by 1938. In 1937 this example of a TA7 was built for John Welch & Sons of Chinley with the style of bodywork so widely used in the cotton piece goods trade of Lancashire. At first sight the 'Badger' of the period was almost indistinguishable from the heavier 'Beaver'.

Below: A busy scene in 1932 with vehicles of the Metropolitan Transport Supply Co. under contract to International Stores. The two vans on the left are on TSQ 'Bull' chassis, while that on the right is a TQ1 'Buffalo', both models being rated as either 6- or 6/7-tonners. Note that the driver of the centre vehicle is in uniform with a peaked cap, and that the van bodies are of heavy double-skinned construction, possibly with insulation for a certain level of temperature control.

Above: Tennant Brothers of Sheffield were the operators of this TQ1 'Buffalo' pictured outside one of their more modern houses in May 1931. Finished in maroon with a light red bonnet, the cab side is adorned with the fine coat-of-arms of the City of Sheffield. The load of casks and crates is retained only by the stanchions and chains which might be sufficient for large casks, so the smaller crates are stacked against the head-board, with the outer edges resting on the rave to assist with stability.

Left: Identification of some models in the 'Badger/Beaver/Buffalo' series is sometimes difficult because of their similarity in appearance, especially when they're not badged! This is in fact a TQ1 'Buffalo', but what is not immediately apparent is that this vehicle had a single-tyred trailing axle fitted for extra capacity. The vehicle is one of the large fleet of Leylands in use by H. Viney & Co., many of which operated with drawbar trailers, and were regularly to be seen at work around Preston docks in the days before nationalisation.

Right: An early morning scene in Manchester fish market as one of Charles Alexander's lorries is relieved of its load of fish which has just journeyed down from Aberdeen. The vehicle is a Kingston-built SKG2 model 'Cub', which has a non-Leyland cab. It is perhaps surprising to find that a small vehicle such as this was used for long distance transport, but the answer might lie in the fact that the machine was light enough to quality for a legal speed of 30mph, against the 20mph limit which applied to vehicles over two and a half tons unladen at the time.

Below: Crosse & Blackwell were a company famous for their sauces and pickles, and they operated quite a number of Leylands in their fleet. In this view of the yard at Bermondsey, taken in October 1932, we have a good view of a 1931 TQ 'Buffalo' with distinctive integral van bodywork, reminiscent of the boarded tilt type, but with a rail round the roof for the empties. In the background can be seen the cooperage, for the repair of the casks used for the imported fruits and spices used in the production process. Immediately behind the van, near the horse-drawn vehicle, are stacked the earthenware jars and glass bottles in which the pickles were packed for delivery, in the days before the modern trend of vacuum packed small glass jars.

Left: Before the advent of road tankers collecting milk in bulk from dairy farms, milk was carried in churns, being collected from a pick-up point along the road, or by the farmer making his own delivery to the creamery. T.H. Bladon & Sons was one of the many haulage contractors who carried out the 365-days-a-year task of collecting the milk in order to keep the local creamery supplied with fresh supplies. The Manchester-based contractor used several Leylands on contract to Harrison's creamery at Chorlton-cum-Hardy, and this pair of Cubs were among them. The older vehicle certainly seems to have seen plenty of action. It is an SKPZ2 model passenger chassis adapted for this particular use because of its lower chassis compared with the standard lorry model.

Right: London Transport operated many 'Cub' models during the 1930s, both lorries and buses, including the revolutionary rear-engined type. This SKZ1 model was supplied in July 1936 for the service department based at Rye Lane, Peckham, and has a special high dropside body for the security of tramway bogie wheels carried to overhaul works. Note the safety side guards which are fitted for the protection of tramway maintenance workers, much of their work being carried out at night.

Below: In 1935 the 'Cub' underwent certain changes which includes new engines, and a redesigned style of radiator which was more in keeping with the heavier Leyland models. An example of the new range is this SKZ1 type in service with the Hull-based cement manufacturer G & T Earle, and is pictured ready to unload its load of bagged cement at Hull docks on a sunny September day in 1936.

Left: The 'Cub' was first introduced in 1930 as a new lightweight model aimed at the medium weight truck market from around two tons payload. In the following year sixwheel models were available, and from 1933 a diesel engine was offered. A redesigned Z range followed in 1935, which was readily identified by a change of radiator shape. This 1937 example is a KZDX1 sixwheeler rated for a six ton payload, operated by Allan Simpson Ltd, a small contractor based in the City of London and operating a fleet of fifteen vehicles.

Below: Although a posed photograph, this view of an SK3 removal van does capture some of the genteel world of a middle-class residence in a leafy part of Surrey, in the peaceful days before World War 2. Bentalls of Kingston are an old-established departmental furnishing store in Kingston-on-Thames, a little more than a mile from the Ham works where the Leyland 'Cub' was built.

Above: A bright sunny day in June 1937 sees this sixwheel 'Cub' in the colours of Brooke Bond Tea outside the company warehouse, close to the Minories on the east side of the City of London. A much earlier TSC 'Beaver' with trilex wheels is being loaded in the background. The vehicles were on contract from H. Hamilton Jardine, whose premises were a few miles south of the river at Clapham. There was still a large number of horsedrawn vans at work in our cities at this time, and they were well used to mixing with the motor variety. This example is left without the wheel being scotched or tied, and the reins are left hanging loose.

Right: In 1937 a new model was added to the range, named the 'Lynx' it really was a reworked 'Cub' but with the scuttle positioned further forward, it was said to be of the 'semi-forward' type. As with the 'Cub', the 'Lynx' was built at Kingston, and the model designation was DZ. Only a few miles from the factory, at Chiswick, W. Ashby & Sons, a firm of pea and potato merchants, operated this example, and it is seen here loading vegetables at the now defunct Brentford wholesale vegetable market in 1938.

Above: London-based meat hauliers T.M. Fairclough ran a very mixed fleet of vehicles in pre-nationalisation days, and Leylands featured large among the wide assortment of types operated. They ordered so many 'Beavers' during the mid 1930s that the fact was given prominence in Leyland advertising of the period. This group of three TC9 Beaver-and-trailer outfits has posed for the camera, the slow shutter speed used making the cyclist appear blurred.

Below: In this 1939 photograph we see a pair of DZ1 'Lynx' models in the colours of the famous Andrews Liver Salts, and we might be forgiven for asking if they were 'regular' Leyland users! This model was aimed at the middleweight market of a five tonner able to travel at 30mph, but it faced strong competition from the light and fast US-style trucks which were selling well in the period immediately before the war.

Right: This 1936 model TSC10 'Beaver' for ICI must have been for something hazardous, for it is of all-metal construction, with cab and bodywork by Duramin Ltd. Unusually, the back of the alloy cab is protected by a steel fire-screen, the special shaped fuel tank has been mounted high up between cab and firescreen, and guard rails span the length between the wheels. Note also the double row front bumper, which is more cosmetic than practical.

Left: E. Coar & Sons was a small haulage contractor based at Preston, and ran several Leylands. This example in their fleet, was a 'Beaver Six' long wheelbase model TSC9, delivered in the Autumn of 1936. It bears all the signs of a typical 1930s general haulage vehicle, from the fixed starting handle, single windscreen wiper, illuminated headboard, railed sheetrack, high headboard and dropside body, to the 12 foot trailer.

Right: Mr. and Mrs. Kime operated a small haulage business based at Wrangle near Boston, Lincs, during the 1930s. In 1938 they registered as a limited company, and the fleet stood at 15 vehicles including two articulated outfits. The 'Lynx' DZ tractor unit pictured here in Boston docks must have been one of the latter. Judging by the close-coupled effect of the trailer it was a Carrimore, for they had developed a patent turntable coupling which had radial slots along which the trailer kingpin could slide as the vehicle turned. This design provided for a smaller gap between tractor and trailer without the risk of collision when turning.

Left: From 1935 the 'Beaver' range was divided into the 'Beaver Four' and 'Beaver Six' types according to whether a four- or six-cylinder engine was fitted. This example for the Burtonwood Brewery of Warrington, was a TSC8 'Beaver Six' with a capacity of around 7¼ tons. Because of the additional option of petrol or oil engines, the unladen weight varied by about 5 cwts, as, in turn, did the payload. The first class paintwork on this 1936 example is worthy of note.

Right: The scene is probably Liverpool docks with the driver tidying up his roping of a considerable load of timber. The three cross ties appear to have little chance of retaining such a high load on a corner, so more faith must be attached to the three vertical stakes, although it is not clear from the photograph how they were secured. The TSC9 'Beaver' has a non-standard cab, and the unusual feature of a single piece, opening windscreen. In retrospect it seems incredible that no nearside mirror was thought necessary, even in 1937.

Left: Springfield Carriers was already an established contractor when it was incorporated in December 1930 under the direction of R.H. Booker. Operating about 25 vehicles and almost as many trailers, the all-Leyland fleet was regularly seen in the Manchester/Liverpool area carrying all types of traffic. In this 1938 photograph they achieved publicity when a TSC8 'Beaver' and trailer had police escort through the Mersey Tunnel with these overwidth steel fabrications. The swinging hurricane lamps are used as marker lights on the load, while the apparent double sidelights are merely double images caused by a double exposure. A nice period piece is the police Royal Enfield combination, with the traffic officer looking decidedly apprehensive about his position in the sidecar!

Right: The elegant lines of the Leyland cab applied to the TSC18 model 'Beaver', are accentuated by the lining-out on this example. Supplied to Andrew Hepburn & Co. of Coatbridge, this machine joined other Leylands already in service with this Scottish haulier.

Below: An historical scene for Liverpudlians, is this 1958 view taken at Canada dock with the Canadian Pacific ship 'Empress of England' berthed in the background. It's not clear whether the Harrison Bros' TSC18 'Beaver' is loading or unloading on this wet October day, but the evocative atmosphere is a reminder of what Liverpool used to be...

Above: Measuring stick in hand, the driver of this early model TSW1 'Hippo' dips the tank compartment to prove it is empty after making the delivery. The vehicle is finished in the livery of Trent Oil Products, whose bulk storage facility was at Gunness Wharf on the River Trent close to Scunthorpe. The T.O.P. pump is located outside the Conisboro' Tyre Re-Rubbering Works, which presumably is a tyre dealer engaged in retreading tyres. With the pump positioned off the road, it was necessary for a swinging boom to be provided so that vehicles could be fuelled while standing at the kerbside.

Right: The 1930s saw attempts at increasing the payload of vehicles following the revised regulations created by the Road Traffic Act 1930, and these usually involved fitting additional axles, normally at the rear. A trend then began for an additional axle to be positioned forward of the rear axle, and to arrange for it to be steered, so as to cut down on tyre scrub. So the twin-steering 'Beaver' came about as a Leyland production option with carrying capacity raised to $9^1/2$ tons. An early example was supplied to Pat Collins who ran Talbot Garage & Transport Co based near Kidderminster, with vehicles operating on contract to Baldwins Ltd, the steel sheet producers at Stourport-on-Severn.

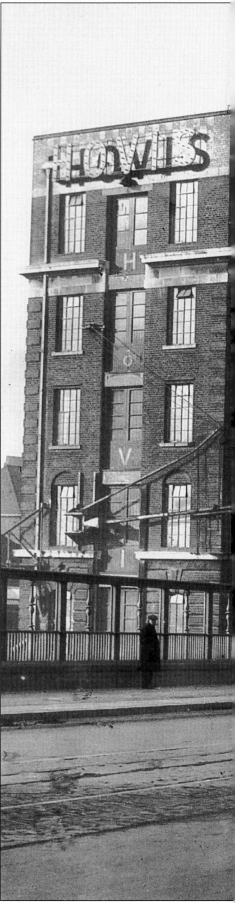

Above: Following the success of the twin-steering 'Beaver' TEC1 model, it was decided that there was sufficient demand for the type to receive an update. The new TEC3 model was given this rather startling new cab design and named 'Steer' in the Leyland menagerie, and some thought the new cab might be extended through the whole range. This particular vehicle, chassis No. 16760, was prepared for the 1937 Commercial Motor Show and was shown in the livery of Blythe & Berwick of Bradford, a very old established Leyland user.

Below: In the field of long distance transport of prewar days, there was a handful of fleets which still stir emotive thoughts amongst older transport men. Youngs Express, Fisher Renwick, General Roadways, Holdsworth & Hanson and Bouts Brothers are eminent in those memories of the fine fleets which used to ply the major routes of the country, right up to the days of nationalisation. The Bouts vehicles were certainly outstanding in all-over white, as seen in this photograph of an early TSW1 'Hippo'. Their main business was in the cotton trade, originating from the Lancashire mills, but return loads could be almost anything which fitted in with the regular trunk operation of the company. With an expanding demand for fresh fruit and vegetables, a regular return service was possible with this perishable traffic from the growers in the Lea Valley area just north of London.

Right: The Hovis clock stands at 11.45 on a January morning, in 1930, as their TSW1 'Hippo' moves southwards over Vauxhall Bridge, straddling the cobbled surface carrying the LCC tram tracks. The massive flour mill forms the backdrop to the scene, and on the riverside part of the building can been seen the suction hoses used to unload grain from the bulk barges brought up the River Thames. This 'Hippo' was soon to be joined by the first diesel-engined Leyland – the 12 ton sixwheel 'Rhino'.

Left: This must have been the longest wheelbase 'Hippo' produced, and the offside frame rail looks so bare with no spare wheel, fuel tank or toolbox. The footbrake servo is visible just behind the cab step, and toward the rear the frame is swept up over the two-spring, double-drive, overhead worm bogie. Of interest is the fitted sheet used on this TSW4 model for A. Clay of Sheerness, for it is supported by a series of hoops and can be rolled forward if necessary.

Above: The gold cockerel of Courage, brewers of Alton and London, stands proudly on this magnificent example of an 18ft 10in wheelbase TSW9 'Hippo', supplied in July 1936. The vehicle would have been used primarily for the transfer of bulk loads between brewery and distribution depots so loading and unloading was carried out at the rear. For the odd occasion when a retail delivery might be made, a small drop section was provided at the nearside front of the substantial railed body.

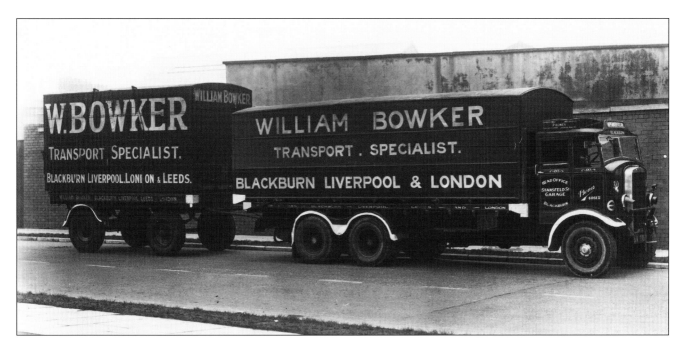

Above: William Bowker established his business in 1919 at Blackburn and later turned to Leylands as he undertook longer distance work. All kinds of traffic was handled on the open goods vehicles, whether it be boxes or boilers, while a couple of containers were held for work demanding a van.

Here we see a 1937 TSW 'Hippo' and trailer with containers loaded during wartime operations. Note the use of white paint on the wings, plus two types of headlamp mask for night running, both mandatory requirements at this period of time.

Above: A slightly different shape is afforded to the alloy cab fitted to this TSW9 model 'Hippo', for both cab and half dropside body were built by Duramin Ltd. The low vantage point used by the photographer gives the vehicle an imposing appearance, which is accentuated by the use of large section tyres, The specification details that 42"x9" tyres were fitted to the front axle, while the rear bogie was carried on 13.50"x20" single giant tyres.

Above: A load of twenty-one rolls of newsprint is a good load for any vehicle, and the Leylands used by Convoys Ltd did it every day. With a wharf at Deptford in southeast London, Convoys did have an advantage over some other newsprint carriers, who had to bring it into Fleet Street from the Medway towns in Kent. Interestingly the vehicle seen here was originally supplied as a sixwheel TSW9D 'Hippo', but has had a second steering axle fitted for greater capacity. It has single tyres on all axles.

Above: In the period just prior to the outbreak of war, Leyland was engaged in producing Home Office heavy pumping units, in preparation for the expected air raids on Britain. Some of these heavy units were mounted on proprietary chassis, while others were carried on trailers. One of the Leyland works transport 'Hippos' is seen delivering six of the pumping units to an undisclosed location in April 1939. This overhead view shows the sliding roof in the driver's cab, which was not usually visible in photographs taken at ground level.

Right: Before most lorry manufacturers had introduced their rigid four axle chassis, a number of three axle vehicles had appeared with second steering axles added. The Leyland 'Octopus' was basically a variation of the proven 'Hippo', and in fact many 'Hippo' chassis were converted by operators to achieve the higher payload possible. As the market for this type of vehicle expanded over the next 30 years or so, all the major manufacturers produced designs, but Leyland and AEC took the major share. This example is one of the three acquired in 1935 by Motor Carriers (Liverpool) Ltd.

Above: It would appear that Henry Shutt Ltd were determined to show that their vehicles were never run at under-capacity! There's not much clearance under the front wings of this TEW9T 'Octopus', as it stands near the Leyland factory with its load of bagged wheat soon after its purchase in March 1935. The body is worthy of note, because is it a dual purpose design.

Above: The most impressive style of rigid eightwheeler is undoubtedly the box van, for it was the epitome of both bulk and strength in motor lorry design. This splendid example of the 'Octopus' TEW8T, was in service with the large wholesale grocers Alfred Button of Uxbridge, who advertised the almost unpronounceable tea on the headboard! Not only are large box vans almost extinct these days, but rigid four axle chassis are rare in haulage, being used mainly for brick and block transport, or tipper work. This vehicle was part of the Nicholls of Brighton fleet.

LEYLAND FIRE ENGINES

Leyland entered the fire engine market in 1909 when an order was received from the Chief Fire Officer of Dublin Fire Brigade for a 'motor fire engine'. Following this vehicles success, Leyland quickly realised that this could be a lucrative market for their quality vehicles, with their 'U' chassis normally being the basis of these machines.

Models were developed to fill all needs, and from 1920 a dedicated range of fire engines was produced, these included motor cycle combination fire engines, a small portable trailer pump, motor pumps and pump escapes of varying size and horsepower rating, turntable ladders and by the late 1930s, aircraft crash tenders.

With the looming threat of a second World War, Leyland built heavy and extra heavy pumping units for the Home Office, as well as over forty fire engines for the armed forces. During the war period TD7 'Titan' and TSC18 'Beaver' chassis were used to mount 100' Merryweather steel turntable ladders.

Post war restrictions meant that Leyland did not pursue the fire engine market. Instead efforts were concentrated on export vehicles and PSVs, however this did not deter one previous Leyland fire engine user, Nottingham Fire Brigade, who adapted the PD2 bus chassis for fire engine use.

In 1958, Leyland, in conjunction with the Chief Fire Officer of Manchester Fire Brigade, developed their first post-war purpose-built fire engine. This was the 'Firemaster' and was technically advanced for its time and revolutionised thinking on fire engine design, but only ten were built for the home market.

Subsequently, all Leyland fire engines have been, and still are built, on adapted goods vehicle chassis.

Above: Leylands' first fire engine was built for Dublin Fire Brigade in 1909. It was based on the then current 'U' type chassis, with four cylinder 48hp engine. It was fitted with a Mather & Platt single stage turbine pump, and first aid fire fighting equipment. it was registered RI 1090. This machine was delivered to Dublin by Mr. Henry Spurrier Senr. During its acceptance trials in Phoenix Park, Dublin, it was called away to a fire at Kingstown which was dealt with promptly. Most early motor fire engines had 'Braidwood' style bodies where the men sat or stood with their backs to the ladder and faced outwards, a very dangerous way to travel at speed. There are numerous accounts of firemen falling off these engines en route to a fire, sometimes with fatal results.

Left: The 'U' type chassis continued as the basis of the early Leyland fire engines with few exceptions. From early 1911 a slightly more powerful engine of 55hp was fitted, a six cylinder engine of 80hp had been available since the Winter of 1910 for fire engine use. The Pump Escape pictured here for Stafford Borough Fire Brigade was new in 1913. Note the Patent 'Non-Skid' solid rubber tyres. This fire engine gave continuous service until 1939 when it was replaced by a new Leyland.

Right: By the spring of 1914 Leyland were also becoming an established exporter of fire engines. Here we see a 'U' type with 55hp engine and Rees-Roturbo pump destined for the Shanghai Fire Department. Leyland had already fulfilled orders from Calcutta and Hobart; many more overseas orders were to follow over the next thirty years.

Left: Of course, London Fire Brigade were the biggest in the country. They purchased their first Leyland in 1911, and by 1919 over fifty Leylands were in use with the brigade. Pictured here is a London standard pattern pump of the time, fitted with four cylinder 55hp engine and Rees-Roturbo pump, capable of pumping 350-500 gallons of water per minute, at 80-100 pounds per square inch pressure – this would provide three good fire-fighting jets. The Rees-Roturbo pump was a two stage turbine type, it became the standard fitment for Leyland pumping appliances, being robust, reliable and relatively easy to operate.

Above: Liverpool Fire Brigade was another major user of Leyland products. The fine body of men, seen here posing for the photographer, look very proud aboard this FE2 Pump Escape. The FE2 had a four cylinder 48hp engine and 500-700 gallon per minute Rees-Roturbo pump. There is no 'bell' fitted to this fire engine, but careful examination of the picture shows a 'gong' fitted below the feet of the officer sitting next to the driver. The gong was a remnant of the horse drawn fire engine era, and some brigades carried on this tradition into the mid 1920's. Another tradition was to name fire engines, this one is called the 'Francis Caldwell'.

Right: By 1920 Leyland had firmly established themselves in the fire engine market. Following many and varied enquiries it was decided to built three sizes of fire engine, thus was born the 'FE' range. The FE1 was the smallest in the range having a four cylinder engine of 36hp and 300-400 gallon per minute pump, the overall length was 17'6". It was ideal for small brigades.

Above: Leyland had been offering a six cylinder 80hp fire engine since the Winter of 1910. Pictured here is a later example for Leeds Fire Brigade. The pump was the usual Rees-Roturbo, but with an output of 700-1000 gallons per minute. It also carried a John Morris 'Ajax' 60' wheeled escape, and first aid hosereel equipment. Unfortunately not many FE3s were made, probably because of their sheer size. One must remember that most brigades were still using fire stations built to house the much smaller horse-drawn fire engines.

Above: Having successfully developed and marketed three sizes of pumping appliance, Leyland then turned their attention to breaking into the Turntable Escape market. Most successful Turntable Escapes were of foreign manufacture. One of these makers, the Carl Metz company, had no British outlet until 1924, when Leyland decided to use this make of ladder. This picture shows the first 'ML' (Metz Ladder), a wooden 85' unit mounted on a Leyland chassis, sold to Manchester Fire Brigade later that year. The engine was the same unit as used in the FE2. The ML and FE types remained available until 1930.

Right: This normal control LTB1 FE with Braidwood style bodywork, and midships pump, was new to the Bristol Fire Brigade in 1931. It could carry a 35' ladder, or a 50' wheeled escape, and was powered by a six cylinder OHC engine. This Leyland is now preserved.

Below: From 1927, Leyland built an envious reputation for their buses and coaches, fitted with the 'T' type over-head camshaft, four- and six-cylinder petrol engines. It was an obvious progression to include this technology in the new fire engine range. Leyland used the LTB1 'Lioness' coach chassis, with shortened wheelbase, for their first pneumatic-tyred fire engines. Here we see a 'half cab' LTB1 FE for Birmingham Fire Brigade, with midships mounted 500-700 gallon per minute pump, and 'New World' body where the crew sat facing inwards. A crew of twelve could be carried this way. Birmingham was to become another large user of Leyland fire engines.

Right: The LTB1 'Lioness' chassis was also used for this fine Metz Turntable Escape, with 500-700 gpm pump. The ladder could reach a height of 85', at the head of which could be fitted a monitor, through which a powerful jet of water could be directed onto a fire. Turntable Escapes, as their name implies, were primarily intended for rescue purposes, the fire fighting capability was a secondary function.

Left: Leyland were producing their side valve engined 'KP' and 'KG' range of 'Cub' vehicles at Kingston, and it was quickly realised that this range could be the basis for a lightweight fire engine. So the FK 'Cub' range was born. Pictured here is an FK1 'Cub' with Braidwood body, for the City of Salford Fire Brigade. It was powered by a six cylinder side valve engine of 27.3hp; the 400 gpm pump was mounted at the rear.

Above: This 'lightweight' FK4 'Cub' Pump Escape for Scarborough Fire Brigade, is fitted with midships-mounted 400gpm pump, and carries a 50' wheeled escape. The open, cross seated bodywork, was to Leylands own design and gained some popularity, as the crew were carried more safely. The engine was the usual 27.3hp six cylinder side valve unit. The FK 'Cub' was well received and found favour with both large and small brigades.

Above: Leyland also produced two sixwheel fire engines, on the 'Cub' KDSX1 chassis. One was for the Borough of Abingdon Fire Brigade. These double drive 'Cubs' had good cross-country capability, which could be enhanced further by the fitting of 'tracks' over the rear wheels. Close scrutiny reveals the tracks in place here, with the vehicle on pre-delivery trials. The power source was the usual 27.3hp six cylinder side valve engine, and the pump is the standard 400gpm type.

Right: In 1935 Leyland improved the Cub range by fitting a six cylinder over-head vale engine of 29.4hp. This required the fitting of a bigger radiator, and along with modifications to the braking system, certainly pepped up the Cubs' performance. These improvements also led to a change in model type. Here we see one of the first FK6 'Cubs' to be built. It was delivered to Glasgow Fire Brigade in 1936, fitted with a Drysdale pump to Glasgow's own specification.

Above: The delivery of Ilfords' new FK6 fire engine was heralded with the headline – 'First Forward Control Fire Engine, Ilford's Novel 60mph Machine'. It was also described rather unkindly as a 'van'. It had a 500gpm pump fitted at the rear, carried a 35' ladder, and could seat up to twelve men. One other item to note in this January 1937 photograph is the 'sliding' crew cab doors.

Right: Not all fire engines were 'all red'. This May 1936 view of a FK7 'Cub' Pump Escape for the City of Coventry Fire Brigade, had the bonnet and mudguards painted red and lined out; the remainder of the open cross seated body was finished in 'Polished Mahogany'. The midships-mounted pump was now rated at 500gpm; the engine was the 29.4hp unit.

Above: An example of progress by the Leyland coachbuilders is this FT4a limousine for Sunderland, supplied in 1938. Up to fifteen coats of paint would be applied by brush, each coat finely flatted down before the next was applied. Lettering was also done by hand, as were coats of arms, if no transfer of this item was available. The mechanical specification was now standardised, the 49.8hp engine and 700-1000gpm Rees-Roturbo pump being the usual units fitted to the FT range.

Left: The further development of the LTB1 types culminated in the 'FT' model. This could only be described as a 'heavy' type of fire engine chassis, and was offered from May 1932. Initially these were mostly fitted with Braidwood style bodywork, but as the type became more popular, other styles of bodywork were fitted to the customer's requirements. This FT2, for the Borough of Heywood Fire Brigade, had a six cylinder 49.8hp engine, and a midships pump of 700-1000gpm capacity. It is seen here, carrying a 60' John Morris 'Ajax' wheeled escape. Note the polished mahogany bodywork which was 'in vogue' from around 1935 until the outbreak of WWII. This machine survived the war and passed into Lancashire County Fire Brigade on its formation in 1948; it gave further service until the early 1960s.

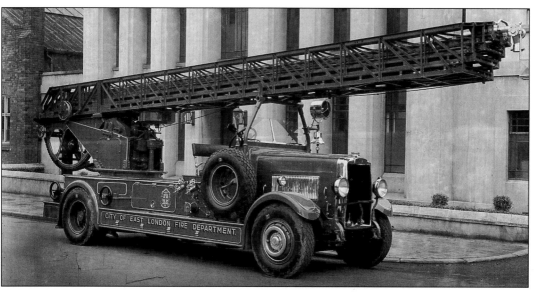

Left: As previously mentioned, some turntable escapes had been built on the LTB1 chassis, the lessons learned from this resulted in the 'TLM' chassis (Turntable ladder Metz). Pictured here is the TLM for the City of Stoke-on-Trent Fire Brigade, with Metz 85' wooden ladder, and Rees-Roturbo 400gpm pump. The engine was a six cylinder 43.5hp unit. This TLM was probably one of the first with polished mahogany bodywork, and remained in front line service until 1954.

Below left: By 1937 Metz were supplying only 'all steel' ladders. This four section 101' ladder on Leyland TLM chassis, was supplied to the City of East London Fire Brigade, Cape Province, South Africa. Once again the power unit was the reliable 49.8hp type, and the pump a 500gpm Rees-Roturbo.

Below middle: Only two forward control FT type were built, this SFT4a for Cork Fire Brigade shows how smart was this design. It was fitted with the usual 49.8hp engine, but in this instance the pump was rated at 800-1000gpm; note the extra suction inlet protruding on the nearside of the radiator. Provision was made to carry a 50' wheeled escape. The spots on the photograph are due to the aging of the negative.

Below: Two TLMs with cabs were built, this one for the Borough of Morecambe and Heysham Fire Brigade, was somewhat unusual in having the ladders and turret painted white. Most ladders were either green or silver. This interesting and unusual TLM is now preserved.

Above: With the impending outbreak of WWII the supply of Metz ladders came to an abrupt end. Modern turntable ladders were in short supply, with a number of outstanding orders which were not able to be completed. To overcome this problem fifteen TLM2a chassis were built and fitted with Merryweather 100' steel ladders; these were delivered to various fire brigades between 1940 and 1941. The TLM pictured is for Liverpool, note the headlamp mask and white mudguard edges which were a wartime requirement.

Above: The 'hybrid' produced by the amalgamation of the 43.5hp 'T' type engine and the FK 'Cub' chassis and components heralded the 'FKT' range; light in weight, with plenty of power. The pump escape pictured, was one of over forty FKTs supplied to the Army and Royal Navy. There were three styles of bodywork supplied to the services, and not all carried wheeled escapes; the pump fitted was a 600-800gpm unit.

Above: The FKT was also built in forward control form. Pictured here, outside Leylands offices prior to delivery in July 1940, is Croydon Fire Brigade's SFKT2 pump escape, complete with 50' Bailey escape and 800-1000gpm pump. This machine, like many pre WWII Leylands, lasted into the early 1960s in service; a fitting tribute to the build quality of the Leyland Motors product.

Above: Post WWII, Leyland decided to discontinue building fire engines. However, the 'Comet' goods vehicle chassis with 100bhp petrol engine found some favour. Fifty were built and were bodied by independent companies as water tender or pump escape. Pictured here is one of the 'rarer' versions. Two of these ECP/1R 'Comet' models were supplied to the West Riding Fire Brigade, fitted with Dutch made 106' Geesink turntable ladders. Both these interesting TLs were acquired out of service by a well known pest control company.

Left: Mention must be made in this chapter of the National Coal Board, Durham & Northern Division Fire & Rescue Brigade. Not only did the Mines Rescue service carry out their main function of rescue, but they were equipped to fight fires in mines. This ECO/1R 'Comet 75', carried a comprehensive range of equipment which included pumps, special clothing, tools, first aid and oxygen breathing apparatus. The bodywork, by Wilson & Stockall of Bury, was finished in cream with red mudguards. Note the Winkworth electric bell mounted on the front bumper.

Right: In 1953 Nottinghamshire Fire Brigade took a great step forward, when they ordered eight diesel powered Leyland PD2/10 'Titan' double deck chassis, on which to built fire engines. The bodywork is reputed to be by Atkey's. With a Dennis pump, each appliance carried 400 gallons of water. Seen here is OAL 1 during pumping trials; the framework protruding from the side of the appliance, is the cradle for the portable pump.

Left: The LAD cabbed 'Comet' did not find favour with U.K. fire brigades, probably because of its awkward access to the cab, but several were used overseas. Here is a 12C/1R 'Comet' for the City of Salisbury Fire Brigade, Rhodesia (Zimbabwe). Note the double width step ring, to assist with gaining access to the cab.

Above: 1958 saw the introduction of the purpose built Leyland 'Firemaster', it revolutionised the thinking on fire engine design. It had a front mounted Sigmund 900gpm pump, under-floor O.600 engine and semi automatic 2 pedal transmission driving a two speed rear axle. Manchester Fire Brigade operated four 'Firemasters', the first, WXJ 286, had bodywork by Carmichaels of Worcester, and carried a 50' wheeled escape. This picture, with its locker doors open, shows the equipment stowage and the bottom locker in front of the rear wheel clearly shows the position of the side mounted radiator. A total of only ten Firemasters were built.

Right: The next stage in the development of the firemaster chassis was a Turntable Ladder variant; two were built. The one pictured here was built for the County Borough of Wolverhampton Fire Brigade, and it had a 500gpm front mounted Coventry Climax pump. The ladder, a 100' hydraulic type by the German manufacturer Magirus, was supplied through David Haydon, as was the bodywork.

Above: The 'Freightline Beaver' had some success as a fire engine chassis, both at home and abroad. This Ergomatic cabbed 16BT2R 'Beaver' for the City of Plymouth Fire Brigade, was designed to replace the need for pump escape, pump and water tender. It was designated as a 'Multi Purpose' appliance, carrying 400 gallons of water, 46' ladder, and a Gwynne Multi Pressure pump. Bodywork was by Carmichaels of Worcester. This appliance was new in 1968 and remained in service until 1984.

Below: The City of Plymouth Fire Brigades' new Ergomatic cabbed 16BT1R 'Beaver' hydraulic platform is drawn up alongside the previously described 'Multi Purpose' appliance in this 1968 view on the forecourt of the Brigades headquarters station. The booms on the hydraulic platform were by Simon Engineering and could reach a height of 65', bodywork again being by Carmichaels of Worcester. This vehicle replaced the Brigades' twenty-six year old Leyland turntable ladder.

ON A WAR FOOTING

By the latter part of the 1930s the international political climate was deteriorating and impending signals of another World War had prompted a start, in 1938, on the building of a special MOS 'shadow' factory at Leyland for the production of military hardware.

At the beginning of the Second World War, the Hippo was modified for government use, and a power unit for the Matilda tank was being developed and tested. Two of these 95-hp powerplants being installed in what became the Mk IIA version of the Matilda III infantry tank built by Vulcan Foundry of Warrington. Leyland Motors Ltd was also engaged in machining, and assembling tank units for other wartime vehicle manufacturers. After Dunkirk, the Ministry of Aircraft Production required more capacity for bomb production, so part of the foundry production was set aside for this use.

In 1941, Leyland delivered its first tank, and by June, the first Churchill tank had been assembled, making Leyland a major producer of both cruiser and infantry tanks. By 1941 500 Covenanters and 754 Churchill tanks had been produced. The former however, was never to see combat as intractable problems with engine cooling resulted in the model being used as a training vehicle only – a role in which it proved invaluable. The Churchill was to prove itself a reliable machine and variants remained in service until 1952. Its good design also led to it being adopted as the platform for a number of specialist roles with the Royal Engineers.

In November 1941 Leyland accepted responsibility for the design of the Centaur cruiser tank and the reliability and success of this vehicle resulted in the company subsequently being chosen to head the Cromwell cruiser tank development and production programme, with the first examples coming off the production line in January 1943. The Cromwell hull was subsequently to be used for the

Below: A peacetime photograph of war preparations; early models of the WLW 'Retriever' ready for delivery outside the Leyland building in 1936. The building nearest the camera in this Thurston Road view, housed the canteen with the drawing office above.

basis of a new up-gunned cruiser tank fitted with a 17-pdr gun capable of challenging the German armour. This was the famous Comet tank. Design work commenced at Leyland in July 1943 and the first production units were reaching the Army's regiments by September 1944. It proved itself a superb machine, combining excellent main armament with cross-country speed and agility, and was to remain in British service up to 1958.

In 1945 production of the Hippo Mark 2 amounted to 40 units per week. In this same year preparations were in progress to return to peace time production of both passenger and goods vehicles. However, Leyland Motors Ltd were not quite finished with the military and its requirements, for, experimental projects apart, production was to commence in the 1950s at that same special factory with the 'Martian' range of heavy cargo, artillery tractor and recovery vehicles.

With the experience gained during the period of war production, the company faced the future with improved designs for new chassis, engines, cabs and running units. The motor industry faced many problems including shortages of raw materials, enemy damage to property, workers to be retrained, new tooling required and a government keen to foster major efforts in exports.

Above: The TE1 'Terrier' was produced for military and cross-country applications, beginning in 1928 and running through in batches until 1940. The heavier WLW 'Retriever' came into production in 1936, with the last chassis being delivered in 1940. This example was produced in 1937 and is seen being checked as the wheels are run up an 18 inch ramp. A take-off shaft on the nearside, suggests that this chassis was destined for searchlight work.

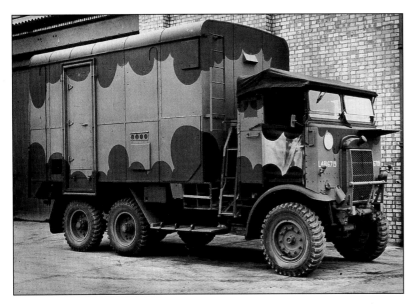

Left: A great variety of bodies were fitted to the WLW 'Retriever' range, these including general service, machinery, breakdown, derrick, wireless, searchlight, Coles Crane, gun mounts plus others. The vehicle depicted here being a workshop variant. The 'L' prefix to the army registration number denotes that it is a truck, 1-ton and over.

Right: An early model 'Retriever' fitted with a general service body. Classified by the War Office in the medium type, 3-ton, 6x4 range, the WLW 'Retriever' was fitted with a 5895cc ohv engine producing 73bhp with power being fed to the rear wheels by a 4-speed gearbox and 2-speed auxiliary box. Tyres were 9.00 x 20.

Left: Later WLW 'Retriever' models did not have the luxury of the fine Leyland aluminium radiator shell: this style with a wire-mesh grille and a crash bar was deemed adequate. This March 1945 photograph details this example as a 'pontoon body, No. 5, Mk1, raft unit'. All told 6866 WLW chassis were built between 1936 and 1942.

Right: One unusual application of the WLW 'Retriever' was this lightly armoured anti-aircraft cannon, here shown with the driver's cab armoured panels folded down. In this design the cannon cannot be rotated, and a tubular cross-bar is fitted which is designed to prevent the cannon being aimed too low. This example is lettered for working with the Ministry of Aircraft Production, for the defence of wartime shadow factories.

Left: The most famous 'Retriever' of all – General, later Field Marshall Montgomery's personal caravan, seen coming ashore from a tank landing craft 'somewhere in the Middle East theatre of operations' – as they used to say.

Right: The wartime military version of the TSW 'Hippo' was designated WSW17, and production totalled almost 3000 in the period 1940-1947. This 1941 photograph shows one of the batch of 330 which went to the Royal Army Service Corps at the beginning of the war, all being fitted with the general service type of bodywork shown here.

Right: A rather indistinct wartime photograph showing vehicles undergoing sea trials in rather choppy conditions 'somewhere in England'. The first vehicle appears to be a four wheel ERF, followed by a military version of the 'Lynx'.

Right: A Leyland production A27M 'Cromwell' tank photographed in May 1944; there were few photographs taken of these during the war period. A top speed of 40mph was claimed for this tank, making it one of the fastest in production at the time.

Below: As Leyland Motors were allocated the work of tank building in the 39-45 war, it was necessary to rapidly design a vehicle capable of carrying a Crusader medium tank. Three of the Works Transport 'Octopus' were converted for this task by the addition of a third rear axle, thus making for a rather cumbersome five axle rigid seen here. Note that a considerable number of substantial new cross bearers were necessary to take the concentrated weight of the tracked vehicles.

Left: Most military subjects were posed against a neutral background when photographed in wartime, so this view of a production 'Comet' with an identifiable background is unusual. Leyland, March 1944.

Below and left: How the tanks were delivered: by road and rail. A small convoy of Cromwell tanks ready to move off from Leyland Motor's Farrington works in April 1945, with US Mack NM6 trucks towing Crane Trailers' design of tank trailers. Others were ramp-loaded at the local railway yard at Leyland onto bogie flat wagons for consignment to the Channel ports.

Above: A view of the chassis production line at Ham works, Kingston-on-Thames, taken in January 1945, showing 'Lynx' and 'Terrier' chassis nearing completion. It is interesting to note that all the wheels and tyres in view are of civilian type, there being no military pattern tyres at this time. The 'Lynx' was one of the vehicle types available to civilian operators during the war, should the MOWT approval be forthcoming for essential war work.

Right: A carefully posed photograph of a Leyland Works Transport 'Lynx' in wartime garb, being checked out with its load of 'Leyland Centaur' cylinder blocks in May 1944. The vehicle is probably an export model that was unable to be delivered because of the conflict, for it is a left-hand drive machine.

Left and above: Two views of the Leyland tank factory at work in 1944. The official title was the BX factory, and we can see 'Comet' and 'Cromwell' gun turrets being fabricated in the left photograph. A further stage of production shows turrets mounted on tank hulls in the photograph above. It is interesting to note that the turrets were given their T-numbers even before they were mated with the tank hulls.

Right: A small part of the wartime production total: 'Hippo' chassis/cabs awaiting despatch to the bodybuilders from Leyland's South Works in March 1945. It is interesting to note that several of the chassis in the photograph are carrying test weights.

Right: The later design of the WSW17 was referred to as the 'Hippo' Mark 2, to distinguish it from the earlier model which had a soft cab. This later model boasted a Leyland pressed steel cab, a similar type being used on the Interim 'Beaver' just after the war.

This page: In 1945 hostilities ended and with it, intensive military procurement. Apart from trials with the FV1000 ultra-heavy tractor, it was not until late 1953 that Leyland saw significant military production recommence with the first of the new FV1100 family of 6x6 'Martian' heavy tractors. Eventually nearly 1400 examples were to built at Leyland's MOS factory in artillery tractor, cargo tractor and recovery vehicle guises, power being supplied by the Rolls-Royce 'B Series' engine. Off-road performance was impressive, there being significant similarities in the suspension design with that of Scammell as exemplified in the early 1950's Motor Show picture. Examples of the FV1119 recovery vehicle soldiered on in service until the 80s; other variants being the victims of military obsolescence well before this time. The lower view shows a pre-production artillery tractor being put through its paces identifying its cross-country capabilities.

POST-WAR RENAISSANCE

Leyland was quick off the mark in the early post war years in introducing a new range of heavy vehicles. The only visible carry-over from early designs was the Interim 'Beaver' with its military pattern cab, and it was not long before the fresh styling of the new era began to appear on our roads.

First came the forward control range of 'Beaver', 'Hippo', 'Steer' and 'Octopus', soon to be followed by the bonnetted 'Comet' which marked a new departure in Leyland design strategy.

The first few years after the war were marked with shortages of all kinds of raw materials, and Leylands were not immune to these shortages. In addition, the accent was on exports, so the home market was starved of new vehicles on two counts.

The road transport industry was to suffer another kind of problem in the late 1940s, when the turmoil of nationalisation was to sweep the country. Leyland were fortunate in that their products continued to be specified by the new centralised road transport operation, BRS, plus the fact that many of the own-account operators bought from the Lancashire manufacturer.

Another vast change in the industry was brought about by the return to private enterprise of a large part of the BRS fleet in the post 1953 era. As these fledgling private fleets expanded over the next few years, the orders were satisfactory, and revised legislation was gradually moving toward larger vehicles.

Above: The early postwar years saw a wide variety of ex military equipment pressed into service, by a haulage industry bereft of new vehicles. Still sporting its military pattern front tyres, this wartime 'Retriever' saw many years service with Springfield Carriers of Manchester from 1947.

Left: Acquired by Toft Brothers & Tomlinson of Darley Dale in 1947, this ex military 'Hippo' is seen standing on dockside railway tracks at Newcastle-on-Tyne, just after the company had been acquired by BRS in July 1950. It must have been a miserable day, judging by the state of the windscreen, and this is emphasised by the tiny arc swept by the six-inch wiper blade!

Left: The first production run of postwar 'Beaver' chassis were designated as IB or Interim Beaver, and they were supplied with this style of Leyland steel cab of wartime design. This 1946 example joined the predominantly Leyland fleet of Joseph Nall & Co., the Manchester based carrier, which had its origins as far back as 1845. The company were cartage agents for the LMS railway, and operated hundreds of horsedrawn vehicles on railhead distribution work, as well as a sizeable lorry fleet.

Right: The early postwar 'Beaver' models were designated as 12/IB and equipped with the austere pattern wartime cab. This 1946 model in service with H. & R. Ainscough Ltd, flour millers of Burscough, was recabbed later in life with this design by Bowyer Brothers of Congleton. The vehicle was later retired and passed into the world of preservation.

Left: The early April sunshine reflects off the side of this recently delivered 12B/1 'Beaver' of Mitchells & Butlers, as it takes on a load of casks, at the Cape Hill brewery in Birmingham. The vehicle was the first of 34 similar examples to join the brewers fleet during 1947-48. In these early postwar days the effects of pallet loading and metal beer kegs were still far away, and the vehicles retain the age-old methods of wooden casks, chain sided bodies and barrel skids for unloading. One small modern detail, almost grudgingly accepted, is the Lucas trafficator attached to the corner pillar of the body.

Above: The Aberdeen-based fleet of Charles Alexander was closely allied to the fishing industry, so it is fitting that this 12B/1 'Beaver' should be pictured with a load of fish boxes against a background of trawlers. The Alexander fleet, which contained many Leylands, fell into the hands of British Road Services in 1949, but the name reappeared in the middle 1950s following the passing of the 1953 Transport Act.

Right: The truly impressive bulk of the flour mill at the Royal London Docks, tends to dwarf this pair of Leylands in service with the Co-operative Wholesale Society. On the left is a model 22.O/1 'Octopus' parked in front of the trailer drawn by a 12.B/1 'Beaver'. The CWS employed a very large fleet of vehicles of all types during the heyday of the co-operative movement, with the majority being centred on Manchester and London.

Right: In the era of nationalisation following the Transport Act, 1947, British Railways ran an enormous fleet of motor vehicles, and while the majority were engaged on railhead distribution, there were some specialist vehicles for other tasks. This Leyland 'Beaver' 12B/1 drawbar outfit of the Southern Region, was based at Nine Elms depot just south of the River Thames, and was regularly seen travelling empty to the Royal group of docks to load imported meat in these insulated containers. This April 1957 view shows the riverside facilities at Nine Elms, with Vauxhall Bridge in the background.

Left: Under the nationalisation of private road transport operators, British Road Services brought together the operations of several large hauliers specialising in meat transport. Vehicles from the fleets of Pickfords, T.M. Fairclough, Matthews & Co., Hays Wharf, E. Wells & Son and others, were grouped together to form the Meat Cartage Service within the Special Traffics division. For handling the then standard insulated meat containers or lift vans, a large fleet of 'Beaver' flats were employed, often with drawbar trailers. This photograph shows one such outfit passing through Barnet in April 1953; note the statutory attendant, or trailer man, in the cab.

Right: In the mid 1950s the Ross Group was a large operator of Leylands, as this 1955 photo-graph shows, with 'Beaver' and 'Comet' rigids lining the quayside. The scene is most likely to be Grimsby, for Ross were in the business of fish, and the insulated containers were essential for the temperature control of the loads as they were sped across the country.

Above & Left: For the transport enthusiast of Liverpool, names such as George Davis, Guinness, Bibby, Tate & Lyle and Jarvis Robinson can bring a tear to the eye. In the 1950s, when Liverpool docks was a busy place, the 'bottom road' witnessed a continuous stream of lorries of all types, many in the colours of the fleets mentioned. The last-named, known locally as JRT, ran a large number of Leylands, with the drawbar tractor being favoured for dock work. The company, Jarvis Robinson Transport Ltd was formed in March 1920, and in the 1930s an A-licence was granted for 25 vehicles and 56 trailers. An example of the prewar fleet is the TSC10 ballast tractor shown with two trailers, one of the rarities of dockland transport operations. The second photograph shows an example of postwar years, with a pair of 12B/7 'Beaver' ballast tractors emerging from the Mersey Tunnel with ships' propellers, destined for the liner 'Franconia' in March 1951.

Left: The 'Comet' was designed and produced at a time when the country was desperate for exports in the aftermath of war. But the model proved a great success in the home market. The Cement Marketing Company was the largest C-licence purchaser, taking several hundred into their fleet. The ECO2/1R was the basic model which sold so well in the first few years of production, and the fresh styling of the front end came as a complete surprise when the range was announced in 1947, and named after the war-time military tank produced at Leyland. The 'Comet' was also popular in miniature, thanks to a remarkably good Dinky Supertoys model of the type.

Right: R & J Strang of Airdrie were the operators of this fine example of the 'Comet', which joined the fleet of around a dozen vehicles in 1949. The driver must have been very proud of his vehicle, for he has added a number of aluminium strips to accentuate the distinctive lines of the front end.

Left: The articulated version of the 'Comet' was almost as popular as its rigid forebear, and a considerable number were to join the fleets of haulage and own-account operators during the 1950s. This 1955 photograph shows an ECO/3R model tractor in use by flour millers, Reads of Norwich, with a bulk hopper type of semi-trailer.

Above: The articulated tractor version of the 'Comet' was most often fitted with the Scammell type of automatic coupling, thus providing a quicker exchange of trailers. This example, operated by Holdens of Manchester, a company which regularly ran a London trunk service, is seen in Long Acre on the north side of Covent Garden market.

Right: A forward control version of the 'Comet' was available from 1953, and embodied a steel cab by Bonallack. This example was operated by Fordath Engineering Co. Ltd of West Bromwich and carried a three-compartment tank by Thompson Brothers of Bilston.

Above: Birkenhead docks in 1958 with a ECO2/5R 'Comet' articulated vehicle operated by Hodkinson Haulage Ltd. This company was such a regular user of the docks that it equipped its vehicles with two-way radio so that rapid contact could be made between drivers and the depot at nearby Atherton.

Below: Taken in May 1956, this photograph shows the access to the South Works of Leyland Motors Ltd, with a works transport ECOS2/1R 'Comet' passing through the gate. The police lodge on the right still stands in the yard of what is now the British Commercial Vehicle Museum.

Below: Another version of a forward control cab on a 'Comet' is this ECOS2/1R example with Boalloy cab, seen working on opencast coal mining by Burnett & Hallamshire of Sheffield in 1955.

Above: The Davis Brothers group of companies had included a large number of Leylands in their fleet, right from the early days of buying BRS units. The early 'Beavers' were gradually replaced by forward control 'Comets', and in this photograph, one of associate company C. Bristow Ltd is being sheeted and roped in Liverpool docks, as the Canadian Pacific liner 'Empress of Scotland' turns in the Mersey as she embarks on her voyage across the Atlantic to the St. Lawrence Seaway in 1957.

Above: The Liver Building, with its far-seeing Liver Birds, forms a majestic backdrop to this view of Vauxhall Road, Liverpool, which runs alongside the docks. The 'bottom road' as it was known locally, was always full of lorries fetching and carrying their loads from all parts. Here a 'Comet' of R. Watson Jnr of Selby is parked up while the driver is off to look for a load or to sort the paperwork. Over to the right was where the famous Liverpool Overhead Railway used to run, carried on high – hence the fond nickname of 'dockers umbrella'.

Left: Pictured bumping over one of the old railway (un) level crossings, is this 'Comet' truck-mixer operated by Keirby & Perry Ltd of Blackpool. The concept of moving concrete in a rotating drum had been pioneered in the 1930s, but it was not until the postwar housing boom came along that the system really took off. The demand for the timed delivery of ready mixed concrete for immediate use, led to specially designed chassis capable of carrying the concentrated weight at the rear end, often being delivered over unmade surfaces.

Below: In the early postwar years Frank Bustard pioneered the roll-on, roll-off (RoRo) vehicle ferry, by adapting wartime landing craft for his Transport Ferry Service, operating between Preston and Larne. The next generation of vessels were purpose-built ships, of which Empire Cedric was one. This photograph is dated September 1948, and show a 'Hippo' of Pilkington Glass being reversed down the linkspan into the cargo deck.

Above: An early postwar production, this 1947 19H/1 model 'Hippo', is pictured passing through Newbury in 1951. Lettered for Kearley & Tonge, a large firm of wholesale grocery and provision merchants, the vehicle was owned by Metropolitan Transport Supply Co., who operated the vehicle under contract.

Above: The 19H/3 model 'Hippo' was a 17ft 9in wheelbase chassis designed for tipper operation, and this example was operated by Stuart Macey Ltd of High Wycombe, and was photographed passing through Hitchin, Herts in 1953. For some unexplained reason, the vehicle bears a Londonderry registration, a rare sight on the mainland, until the current fad of using N. Ireland number plates by some poseurs to disguise the age of their vehicles.

Left: The 'Hippo' of the 1950s was a distinguished looking vehicle when turned out in a fine livery, this example in the colours of Charles Brown, the London flour millers. The full load of flour has been stacked in a special way to aid stability, with the small front section of the body giving form to the double stacking along the remainder of the load. A few strands of rope around the rear of the load was found satisfactory through experience, and the heavy, rough texture of the bags had great friction and bound together well. Note the additional front bumper which is attached to the front of the chassis frame members, far more effective than the original fitting lower down.

Below: The 19H/3 model 'Hippo' tipper shown here was supplied new to Hilton Gravel in 1953. The whole operation of gravel extraction is shown here, from the dredger barge on the flooded quarry, by means of the bucket grab to the screening plant, up the conveyor to the stacking area, thence by this little Fordson-based Muir Hill loading shovel to the tipper for delivery. Note that the loading shovel is of the old mechanical wire rope design, no hydraulics being used.

Above: By the early 1950s much of the long distance road transport system was in the hands of British Road Services. They, of course, inherited a large number of Leyland lorries through their takeovers of individual companies, and they continued to purchase considerable numbers as time passed. The 'Octopus' featured heavily in the depot fleets engaged on long distance work, and in this photograph we see a 1950 machine based at Blackburn, loading boxed fruit in Long Acre. The canopy of Covent Garden L.T. station projects on the right, and the main part of the fruit and vegetable market could be found down the first turning on the right.

Right: One variation of the 19H 'Hippo' model chassis was its application as the basis for mobile cranes, as this example shows. Here, Smith of Rodley have mounted the crane base directly over the rear bogie, added stabilising legs front and rear, and adapted the Leyland cab to provide clearance for the lattice girder jib during transit. Later regulations have demanded that the jib be hinged back or removed altogether for on-road movements, in order to avoid accidents with such long forward projections. This example was operated by Manchester Corporation Water Works.

Left: An early postwar example of the 22.O/1 'Octopus' seen here in October 1947 in the livery of Risdon Semper & Co., a company within the R. Davis (Haulage) Ltd group. The driver has added a couple of decorations to the radiator grille: with the title of Highland Piper he was no doubt from north of the border.

Above: Some of the most impressive of BRS vehicles in the mid-fifties, were the high capacity box vans which had come to them by way of companies such as Fisher Renwick, Bouts-Tillotson and Holdsworth & Hanson. Pictured passing through Baldock in August 1955, is this 'Octopus' 24.O/4 box van operating from unit 50D Huddersfield, West Riding Parcels, and originally in the fleet of Holdsworth & Hanson. Vehicles of this open top design were used for the Yorkshire woollen trade, where much of the loading might be from the upper storeys of mills.

Above: Many of the photographs in the BCVM's Archive, provide a record of social and industrial history, as a bonus to the vehicles which are the main subject. This photograph of Joseph Crosfield's premises in Warrington in July 1957, shows an 'Octopus' of John J. Potts in its working environment. The vehicle has just loaded boxes of margarine further down the street, and has pulled clear of the works railway line, to sheet and rope. Beside the parked Austin A40 is the fading sign of a wartime Emergency Water Supply, and next comes an old style cast iron road sign with warning triangle. A little 0-4-0 diesel shunter brings in some railway vans, while beyond the glazed walkway a couple of lorries are taking on loads.

Left: The rigid eightwheeler is an impressive vehicle as a boxvan, but doubly so when it has received the streamlined treatment. In 1953 the producers of Domestos decided that valuable publicity could be gained from a vehicle with pleasant lines, coupled with a bold advertising message. Modern Coachcraft was responsible for the bodywork.

Above: Unfortunately a posed photograph, but attractive all the same, is this shot of a 1955 24.O/4 'Octopus' of Spillers Ltd parked in the slow lane of the Mersey Tunnel. The photographer's car – a 1954 Vauxhall 'Cresta' – plus the Land-Rover of the Tunnel Police, are parked clear of the lorry, while a couple of ghostly policemen hold back the light traffic of the day, while the photographer makes his exposure.

Right: Another famous name in Liverpool road transport was that of Kirkdale Haulage Co., an old established business first registered in 1915. During the 1930s they operated a considerable number of steam wagons, which were gradually replaced by IC-engined lorries. Leylands featured in the fleet, and this 1953 22.O/1 'Octopus' seen entering the Mersey Tunnel with a load of BICC cable, was typical of their postwar equipment. A fine array of postwar vehicles can be seen passing through the toll booths in this 1957 view, while the roof of Lime Street station peers round the side of St. George's Hall.

Below: This model 15S3 'Steer' must have been among the last of the type produced when built in 1959, for this style of cab was gradually being replaced through the range at this time. The vehicle, one of 17 similar examples, was operated by London Carriers Ltd, the transport arm of the Phillips Electrical Group, for the movement of household electrical appliances. These vehicles often operated with drawbar trailers, and in their plain mid-green livery with white lettering, were not dissimilar to the BRS parcels fleet of the day. Phillips Electrical also ran a small fleet under the name of Provincial Carriers: these being 'A'-licenced.

Right: As the design of commercial aircraft has improved and their capacity increased, so the demand for larger fuelling vehicles has had to be met. In the immediate postwar period, Thompson Brothers of Bilston designed a range of specialised aircraft fuellers, complete with pumping equipment. There were three designs, all using Leyland running gear and components in their specification. Named after rivers, they were the 'Thames', the 'Tweed', and the 'Tyne'. This 1948 photograph shows the smaller model in service with Shell BP, servicing a Trans Canada Air Lines DC-4M-2 'Skyliner'.

Below: An early delivery of the postwar 15S/1 model 'Steer', was this 50 barrel beer tank for Courage, delivered in 1948. It would appear that the tank is being loaded by the man on the right, who is in control of the hose connected to the filling cock. With the manway lid laid open, the man on the ladder seems to be watching the beer rising inside the tank – a rather crude method of judging when it is full. For some reason the headlamps have been removed from this tanker, so perhaps it was used for internal transport only.

L eyland lorries enjoyed good export sales almost from the inception of vehicle GO FOR EXPORT!

manufacturing at Leyland. The first recorded export sale was for three steam mail vans for Ceylon in 1901.

It was not long before orders were being completed for vehicles for Australia, Argentina, Shanghai, Tasmania and New Zealand. In the days following the Great War, Leyland opened branches in Sydney, Australia and Wellington, New Zealand, in order to boost export orders from this corner of the Empire.

Agencies were set up in other countries such as South Africa and Canada, where a Leyland depot was opened in Toronto. In order to service enquiries from overseas, the London branch at New Kent Road was designated as the export office of the company.

By 1920 the company listed agencies in Brazil, China, Ceylon, Denmark, East Indies, France, Turkey, Greece, Norway, Singapore, Sweden and Siam. The fire engine market was also explored and naturally the main sales were to countries within the Empire and Colonies, with India and Tasmania featuring in early catalogues.

Expansion of the export department continued into the 1930s with a total of four branches in Australia, six in New Zealand, five in Canada, three in India, and one each in South Africa and Greece.

The number of overseas representatives in addition to the branches also increased, with sales outlets in Cairo, Kuala Lumpur, Rotterdam, Tokyo, Bucharest, Penang, Singapore, Madrid and Montevideo, plus five locations in Persia, seven in South Africa and no less than fourteen in New Zealand.

All this activity came to a grinding halt in 1940, but at the end of WWII the government was urging the whole of British industry to go for exports at the expense of the home market. Beset by shortages of many raw materials, the motor industry responded in fine fashion, with Leyland attempting to restore all its pre-war markets as well as opening up new ones.

The fading Empire and the emerging Commonwealth were fertile ground for trade, South America was still buying because of a lack of local production, and the Middle East was seen to have great potential.

The new 'Comet' plus the export versions of 'Beaver', 'Hippo' and 'Buffalo' all took their place in overseas fleets. In Europe the company formed three associate companies for the assembly and sale of chassis in Spain, Holland and Belgium, while in India, Ashok Leyland was formed to handle the assembly and sales in that sub-continent. A similar venture followed later in Iran.

While exports to some of the far-away places such as South America, Australia and New Zealand continued at various levels of activity, the story in Europe was very different. Faced with various forces in the shape of national allegiance, local vehicle manufacturing, changes in currency valuations and strong competition from other vehicle building countries, the near continent of Europe has proved to be a most difficult market over the past 30 years.

Right: The peeling paint of the Motor Repair Works provided a contemporary background for this export 'C' model at work in New Zealand. Dating from the 1920s, the vehicle has been updated by the fitting of pneumatic tyres during its life, but it was still operational in 1939. The Leyland 'Lorries for Loads' sign was often used in publicity material, and is almost a copy of the trademark of the period.

Above: Only a handful of the TR1 model 'Rhino' were built during the 1930s, one for the home market, and the remainder for export. This 1935 example was for the large articulated tankers used in the exploitation of the oil reserves in Iran.

Below: This 1935 TSA9 'Badger' was supplied to the La Plata Reel Cotton Co. who were based at the town of the same name near Buenos Aires. The company specialised in cotton and silk thread, and their 'Elefante' and 'Machete' trade marks are displayed on the side of the obviously locally-built cab.

Top left: Export vehicles often have a sparse look about them because of the lack of rear wings, but this export TA8 'Badger' can boast a practical cab design, with a sleeper compartment and king-size sheet rack. Photographed in 1938, the outfit was operated by A.E. Johnson from New South Wales.

Left: A fine example of a pre-war TSW14 'Hippo' operating in Argentina in 1948, which has just taken on a capacity load of Waldorf toilet tissue! Operated by Jesus Vidal, who had premises in Cordoba and Buenos Aires, the driver was very glad to have a sliding roof in the cab as well as double opening windscreens.

Above: During the 1930s there were attempts at streamlining almost everything, from electric irons to steam locomotives, and trucks received a small share of the treatment. We aren't told how the British & American Oil Co. viewed this attempt at streamlining this 1936 TC14 'Beaver', but with the steering wheel coming halfway up the windshield, one can guess at the reaction of the driver! Operating in Canada, this five-compartment tank semi carries advertising for the company's 'Nevr Nox' petrol – it could only happen in North America.

Below: An early postwar export to Argentina, was this 1948 EB/12L 'Beaver' artic in service with Sanchez & Alonso. This impressive box van was operated over the 300km route between Buenos Aires and Rosario, in the Santa Fe province.

Top: On such a vast continent as Australia, road transport has played the major role in moving the country forward, especially in the areas away from the coast. The movement of sheep is of prime importance to the economy, and this pair of 'Super Hippo' EH/2R drawbar outfits operated in Queensland from 1949.

Above: The Leyland 'Meteor' chassis was virtually an export version of the 'Comet' and at first glance this outfit could pass for a home product with its load of straw. It was actually supplied to a Dutch operator in 1961. Some of the models supplied to Australia embodied AEC engines and running units.

Left: In the days when some factory-built lorry cabs were rather austere and comfortless, many continental operators arranged for chassis to be fitted with specially built units by professional bodybuilders. This Leyland 'Meteor' artic for a Netherlands operator has a pleasing design of cab which the builder has managed to blend with the lower section of the contemporary LAD production.

Above: A 19H/1 'Hippo' tractor heads this heavy haulage outfit, which has loaded a 50 ton item from a ship in a New Zealand port, and prepares to move off. The single ended trailer is connected via a wheeled gooseneck, and a large notice on the trailer announces that the 80 ton capacity transporter by Tapper has hydraulic suspension and steering. Another six wheel tractor acts as a pusher unit at the rear.

Left: The corrugated roofing throws an unwanted shadow over this 12B/1 'Beaver' tanker, as it loads aircraft fuel at the BP depot in Lisbon. Operated by the Aviation Service of BP in Portugal, this 1957 vehicle with locally built cab, ferries supplies the two miles between fuel depot and airport.

Above: How times have changed. In 1955 Leyland was able to export this fine pair of 'Super Hippo' drawbar outfits to B. Andersson in Sweden, who operated under the ASG umbrella.

Below: There is an interesting mix of transport within this 1957 view of Amsterdam, as an ECO/3L 'Comet' artic tanker of Shell heads for its next delivery in the city.

Below: Another difficult export market was Switzerland, which had ample vehicle production of its own. This model 14B/13R 'Beaver' and trailer was in service with A.G. Roth of Basel in 1964. Interesting to note that right-hand drive has been specified, so useful for watching the edges of some mountain roads.

Above: At first sight, one can be forgiven for thinking this was a Schweppes vehicle in Britain, for it is righthand drive, and bears a familiar style of trade-plate. But, it is in fact an export 16S/3 model 'Steer' photographed at the bodybuilders before entering service in South Africa, in August 1958.

Left: This is a special length 'Octopus' model 26OT/2 chassis, one of six built for Shell International in 1968, as the basis for aircraft fuellers in the Far East. When fully laden, the top gear speed was a mere 6mph, and considerable tyre scrub was experienced on the rear bogie as the vehicles manoeuvred around the aircraft. Examples went to Bangkok, Hong Kong and Jakarta, and Gordon Baron worked on the chassis during building at Leyland.

NEW DESIGNS: NEW HORIZONS?

Production at Leyland was still healthy in the early Sixties with good export orders and a home market waiting list for various types of chassis. But this was a period of change – and by the end of the decade much would have altered on the road haulage scene and with Leyland in particular.

In 1964, new Construction and Use Regulations came into being, bringing with them a change that was to see the decline of the rigid eight's prominence on the haulage scene with operators transferring allegiance to the 32-ton gcw artic and its advantages in payload. Such was the preference for the artic configuration that by 1970 Leyland was to drop the eight-wheeler 'Octopus' from its range and it was not for another five years before the type was again listed.

The Ergomatic-cabbed range of lorry chassis was introduced in 1964 and the Leyland-Albion-Dodge (LAD) type cabs sourced from Motor Panels phased out on all models. This cab design was also being introduced into the AEC and Albion model ranges. The Ergomatic tilt cab, designed by Michelotti and built by Joseph Sankey Ltd marked a great step forward, offering far greater comfort and a much improved working environment for the driver. It was readily accepted by the transport industry.

The fortunes of Leyland Motors however, was much influenced by the political and economic thinking of the 1960s, which favoured mergers and conglomerates – the British aircraft industry was already going down this route and now attention turned to the motor industry. The merger of Leyland Motors in 1962 with the ACV Group of companies, and the previous mergers with Albion and Scammell in 1951 and 1955 respectively, having a major effect on the economy

Below: Seen leaving Tower Bridge is this CS3/3R model 'Comet' rigid operated by Mason Brothers (Haulage & Storage) Ltd of Rotherham, with a full load of fish boxes. With an office in central London, Mason Bros vehicles were regularly seen around the fish and produce markets, and they ran a regular nightly trunk between the capital and their native Yorkshire towns.

of the new group. Product rationalisation began to take place with some models being, at first glance, distinguishable only by their constituent company badging although closer inspection would reveal grille variations and other individual manufacturers' characteristics and preferences.

In 1968 the Leyland Motor Group merged with British Motor Holdings to form the British Leyland Motor Corporation (BLMC) and the name of Leyland Motors Ltd was subsumed into the Leyland Vehicles Ltd element of the conglomerate. The commercial vehicle names within the

Corporation now comprising of AEC, Albion, Austin, BMC, Guy, Morris, Scammell, Thornycroft and Leyland itself.

Leyland suffered through being merged with the car manufacturing side of the new, larger business of the Corporation with a resultant pressure on available capital for the research and development needed for continuous improvements to the range of trucks. This was especially vital in respect of the heavier models in the range as the latter part of the decade saw the arrival on the scene of the first high specification tractor units from European manufacturers such as Volvo, Mercedes & DAF, although only a trickle at first, it was a portent of things to come.

In the 1970s rationalisation of the model range increased apace and eventually some of the famous lorry names within the corporation were to disappear in the process, as did BLMC itself when its shares were acquired by the National Enterprise Board at the Government's behest in order to save what was effectively a bankrupt business, the new nationalised company becoming British Leyland Ltd. The corporation was brought down, amongst other reasons, by a lack of investment in new models, tooling and the manufacturing infrastructure – it already having huge borrowings. The heavy vehicles side was still profitable however, but even this was not enough to offset the huge losses of the car making division.

It is now the Nineties and the Leyland name lives on in truck manufacturing – albeit not without a roller-coaster ride or two, but today the focus of this book is to celebrate the achievements of the company that began all those years ago in Herbert Street, Leyland – Leyland Motors Ltd.

Left: This new style of cab originated on the 'Comet' range in 1958, and was gradually extended to the other models. Leyland described it as the 'Vista-Vue all-steel luxury cab', while in the industry it became known as the LAD cab, because it was a Motor Panels production shared by Leyland-Albion-Dodge. The model shown is a 14LWB/1R 'Badger' and was slightly shorter and lighter than the 'Beaver' of the day.

Right: A view of the Mayfield transport cafe on the A6 between Preston and Lancaster, taken in June 1966, finds a trio of Leylands together. On the right is a Freightline 'Badger' alongside a pair of LAD cabbed 'Beaver' articulated tankers, operated by T. Brady & Sons Ltd of Barrow-in-Furness. It was in 1964 that the Ergomatic style of cab was introduced on certain models, the 'Beaver' being the first.

Left: In the early 1960s British Railways, in collaboration with The Pressed Steel Company, produced a revolutionary road/rail system which BR hoped might attract road hauliers to use the rail system for long distance journeys. The basic idea of the system revolved around the use of special semi-trailers, which were equipped with both road and rail wheels, with a pivot mounting to ensure rapid change from one mode of transport to the other. By the time a handful of the trailers had been built and trials began, the use of ISO containers in trainloads was thought a better route to follow, and so the Roadrailer was laid to rest.

Right: An unusual role for a 'Beaver' is this outfit for steel stockholders Gambles of Belfast, which utilises a 14B/12R model rigid coupled to a special turntable bolster, which is carried on a pair of oscillating axles. As the steel sections are loaded as shown, the bolsters of both vehicle and 'dolly' trailer, can turn to accommodate movement.

Left: It had been snowing the night before the Leyland photographer visited James Durrans & Sons yard at Penistone, on the east side of the Pennines in February 1964. He found part of the fleet being prepared ready to set off from Phoenix works, and in this shot he captured a 'Comet', an articulated 'Badger' and a pair of Albions. Note how the vehicles are prepared for the weather, by the use of close-fitting radiator muffs.

Left: In recent years the trend has been for brewery vehicles to display the names of individual beers, for often the cost of painting is debited to the publicity account, and not that of the transport department. This 1960 photograph shows a Leyland 'Comet' tractor, coupled to a dual-compartment stainless steel tank, carried on Tasker running gear. It displays the name of Carling Black Label Canadian Lager, but is in fact operated by Hope & Anchor Brewery, Sheffield. In the early days of 'Comet' production most tractor units were fitted with Scammell type automatic couplings, but in the 1960s the swing was toward fifth wheel couplings as that shown here.

Right: The Bolton-based brewers, Magee, Marshall & Co were Leyland users for very many years, and in 1960 added this model 14SC/14AR 'Super Comet' to their fleet. The vehicle was operated with a drawbar trailer, as were the older ECO 'Comet' models in the fleet, and in that respect the company was almost unique. This outfit appears to be all ready for the road, and the driver happy with the strands of rope around the metal crates, while on the trailer the casks on the roll are secured by means of roping over the folded sheet.

Left: It is in situations such as this, that model identification is not so simple. Both these vehicles, in the livery of Ribbledale Cement of Clitheroe, are in the 'Comet' range, and appear identical from the front. The rigid vehicle nearest the camera is a 'Super Comet' model designation 14SC/11R, whilst the articulated outfit behind is a 'Comet' listed as a CS/14R, both vehicles dating from 1961.

Above: Even before the general swing to articulation during the 1960s, many transport operators favoured the system because of the flexibility of operation provided by being able to switch trailers to meet changes in demand. In brewery operation, the shuttle system of picking up and dropping trailers at either end of a short journey might be beneficial, or the flexibility of using the same tractor with trailers of different types according to trade demands, i.e. for bulk tanks, flats or vans. Here a 'Comet' with three compartment tank of Ind Coope, prepares to leave the Burton-on-Trent brewery on a bright March morning in 1962.

Left: The Blackburn-based haulage contractor William Bowker started business in 1919 and quickly grew into a sizeable fleet of around 35 units by the 1930s. Absorbed into BRS in 1949, the company was re-established after the 1953 Transport Act forced BRS to sell off a large part of its fleet. Many vehicles in the early Bowker fleet were Leylands, and the trend was continued in the post-1953 fleet also. This 1963 photograph shows one of their 'Super Comet' rigids being loaded with fruit for a speedy return to the northwest.

Left: A specialised facet of road haulage is that of livestock transport, which is usually seen and associated with cattle markets and livestock sales. W. Berry of Ribchester near Preston were the operators of this stylish double-deck live-stock truck, based on a 1959 14SC model in the 'Super Comet' range.

Below: From their base at Peterhead, in the far north of Scotland, Sutherlands carried out the longest of refrigerated transport runs. This late 1962 model 'Beaver' is coupled to a 30ft York tandem, with a Thermo-King controlled, insu-lated container. Seen loading at a coldstore, the chilled beef would be delivered to the wholesale market overnight, ready for the retail butchers early next morning.

Right: As mentioned elsewhere, the LAD cab was sourced from Motor Panels and introduced to the Leyland range gradually, it being first fitted to the lighter models. Here we see a couple of Leylands with the LAD cab, but on second glance they are not identical, that on the left is a 'Comet' with the cab set squarely above the front axle, resulting in a short door above the front wing. The vehicle on the right is a 'Beaver' tractor, the cab of which is positioned further forward, allowing a full depth door with the step immediately in front of the wheel. Both vehicles are in the colours of Fleetwood Fish Transport, a reminder of the days when Britain had a fishing industry and Fleetwood was a busy fishing port.

Below: The model name 'Retriever' was first used in 1936 when it was applied to the WLW1 chassis, a cross country sixwheeler for rough terrain and military applications. The WLW model was phased out in 1942, and the 'Retriever' name not used until 1964, when it was revived for a 6x4 chassis designed primarily for tipper/mixer applications. This example is in the orange and black livery of Ready Mixed Concrete, which came to operate a large number of this type.

Above: To some, the LAD cab was downright ugly, while to others it was accepted as being typical for the period. When seen in a Scottish livery, and then on an eightwheel rigid, it is just powerful. The dark blue of Gibbs livery, with just enough lining and decoration to make it tasteful, makes this outfit one of the most attractive ever. You can't get much further northeast than Fraserburgh, yet Gibbs' vehicles are probably appreciated just as much by enthusiasts in London as in Scotland, such is their impact.

Below: Loaded and ready for sheeting, this 1961 model 24.O/12 'Octopus' is operated by A. Wishart & sons of Kirkcaldy on the Firth of Forth, delivering Nairn linoleum. This is another Scottish vehicle turned out in fine style, with duo-red paint and gold colour lettering. Note that the vehicle is complete with a deep chrome bumper, twin spot lamps and front towing eyes.

Opposite page top: It must have been a quiet day in April 1960, when the Leyland photographer caught up with this LAD-cabbed 'Comet' artic, parked up at the edge of the lake at Bowness-in-Windermere. Just visible on the pier is a British Railways totem, for the lake steamers came under the control of BR during the period 1948-1960.

Opposite page bottom: Pictured alongside the Reading abattoir, this sixwheel version of the 'Comet' was in service with Alf Meade Ltd from 1961. Based on the 16ft 11in wheelbase chassis, it was fitted with a 22ft Litex body built by Smith's Delivery Vehicles of Gateshead. It featured polystyrene insulation, a Thermo-King refrigeration unit, and a Smith's 'Market Loader' for the mechanical handling of carcasses when loading/unloading.

Above: This photograph provides a comparison between the new Ergomatic cab and the older LAD production. The two examples are separated by just one registration year, and yet the impression is of a much greater span of time. Both vehicles come from the Furness peninsula – the sixwheel 'Retriever' with its triple deck body from near Ulverston, and John Stable's smaller 'Comet' from Bardsea on the coast.

Right: Much publicity was given to the Ergomatic cab when it was introduced in 1964, the Leyland Journal even putting a cut-out on the front cover, which could be tilted like the real thing. Pictured here is a 16SC/1R model 'Super Comet' based on the long wheelbase chassis, being loaded with bags of animal feed with the help of a powered conveyor, in 1967.

Above: The recent trend of equipping commercial vehicles with their own on-board methods of load handling, has been of enormous benefit to all concerned. By positioning the HIAB hydraulic jib in the centre of the trailer, this driver can safely reach any part of the load and transfer it as required. Halpin Haulage of Bermondsey were the operators of this 'Beaver' with Crane Fruehauf tri-axle semi, and it is seen here with a load of crated glass alongside the Thames just below Tower Bridge.

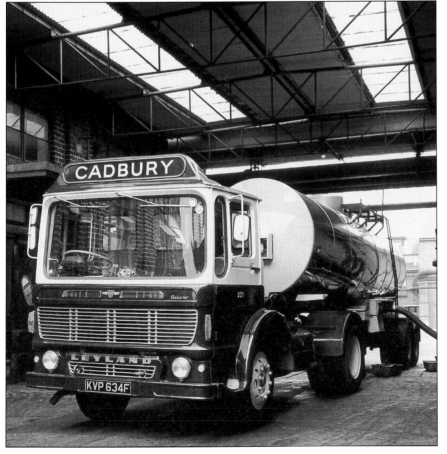

Left: One of the main advantages of road transport, is its reliability as part of the production line of British industry. In this 1967 photograph, a 14BT/18AR 'Beaver' tractor is used to haul bulk loads of hot liquid chocolate between the Cadbury production facility, and the confectionery customer who requires the product to be delivered on time – and not in a solid lump! The well-insulated tank had a load temperature gauge positioned within view of the driver in his cab.

Above: It was a bright and sunny day in October 1967, when the photographer found this model 14BT/28AR 'Beaver' tractor with Boden tandem axle semi-trailer, among the stacks of sawn timber on a Liverpool dockside. The outfit, plated at 28 tons gross, was operated by Pritchard Brothers of Llanerchymedd in Anglesey, and the crew were enjoying a break before loading began.

Left: The move to the bulk handling of animal feeds was well established in the 1960s, but there were still many farmers who did not possess the silos or bulk handling equipment necessary. Burroughes & Stratton of Holt, Norfolk wanted to operate the largest vehicle possible, on economic grounds, but retain the capability of carrying bagged commodities. This 31ft 6in Murfitt trailer with moving bulkhead on tri-axle running gear, could handle either, and with its two-pedal semi-automatic model 14BT/28R 'Beaver' tractor could achieve a 20 ton payload.

Above: Everything in this photograph has a thin coating of white china clay. Visitors to the Par area of Cornwall are soon aware of the main industry, by the huge white mounds near the quarries and the vehicles of Heavy Transport Ltd. In this photograph a model 26OT1/2AR 'Octopus' is seen preparing to tip its load on to a covered conveyor which will transfer the white clay to a ship for export.

Right: The road transport industry in Scotland, has long been renowned for the fine, colourful finish of its lorries, as any visitor knows. Many of the vehicles are given a decoration of tartan and maybe a thistle or two, and the drivers strive to keep them clean in the face of the sometimes atrocious weather of the north. This 'Beaver' with Crane Fruehauf tri-axle semi of Pollock was turned out in this style in 1968, and is seen prior to registration.

Left: Leyland tended to keep the title 'Super' for its export models such as 'Super Beaver' and 'Super Hippo', the exception being the 'Super Comet' which was freely available for the home market. For a very few special orders, the 'Super' export models could be obtained for home delivery and this 'Super Beaver' model EB3/BR in service with Rollon Transport of Little Easton, Derbyshire, is one such example. Although spares for these relatively rare vehicles were no problem, operation could be difficult if trailers were switched, for a 40 ft trailer would render the outfit overlength.

Right: The styling of the 'Super' range gave the appearance of being trapped in a time warp, for it changed very little in outward appearance, with the exposed aluminium radiator shell harking back to the 1940s. Earlier models in the export range had single headlamps mounted on the front bumper, but later examples had twin headlamp clusters within the front wings, as on this EH9/BR 'Hippo' at work in a Derbyshire quarry in 1968.

Left: A glimpse of an export type 'Super' model could occasionally be had in the days when Leyland Motors were testing the fixed-head '500' engine, in almost continuous motorway running. This export 'Super Beaver', loaded to about 20 tons was regularly seen along the M6 on extended trials in 1968.

It was not until the late 1950s, that colour came to be used by the Leyland Publicity department in any volume. It was usually reserved for cover pictures for the Leyland Journal, or for the large poster type illustrations used at vehicle exhibitions.

The BCVMT Archive has produced a small selection of interesting images in colour, to which have been added a few of the hand coloured prints taken from earlier sales catalogues.

LEYLAND LORRIES IN COLOUR

Left: When sifting through Leyland archive material, one is aware of certain early users who made several repeat orders, and were often mentioned in advertisements. Mann, Crossman & Paulin Ltd, brewers of Whitechapel, were users of Leyland steam wagons in about 1903, with others joining the fleet in the next years. They also bought early petrol 'lurries' as Leylands first called them, and this 4 ton model is one example which graced some of the early sales literature.

Left: The City of Westminster was a cherished customer in the early part of the century, for their first three vehicles had been Thornycroft steam wagons. These were joined by three Leyland steam wagons, and the situation watched closely from both Basingstoke and Leyland. Next came an order for some petrol tip wagons, and they were lavishly produced in colour in the company sales catalogues. One important feature of the sale, was the fact that interchangeable bodies were supplied: this refuse collector or a tank for street washing.

Left: Early Leyland Motors sales literature was of generous proportions, covering the whole range of steam wagons, petrol lorries and buses, fire engines and tramcars, and sometimes they were hardbound. One added attraction was to include a selection of full-colour illustrations of which this fire engine is one. This example for Sheffield, was among the first few U-type machines supplied, and was powered by the new, large 80hp, six-cylinder engine.

Right: The town of Leyland can also boast to be the trademark of a famous paint manufacturer, as well as that of a lorry builder. This artist's impression of a 'Cub' 2 ton box van comes from Leyland publicity material of the early 1930s, and although very colourful, the radiator seems to have received some 'artists' licence' in its reproduction.

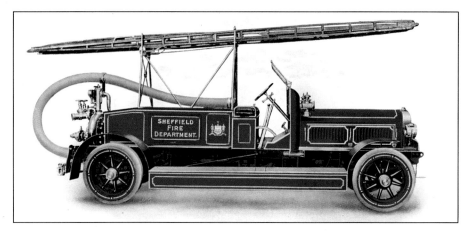

Left: So great was the importance of the RAF machines to the postwar economy of Leyland Motors, that a special brochure was commissioned to publicise the amount of rebuilding carried out. Under the title of 'Pulling the Leyland RAF to pieces – and putting it together again' – the 12 page cut-out type publication, showed a plan view of the chassis, plus photographs of the rebuilding process as well as illustrations of RAF types in service.

Right: Of particular significance to Liverpudlians, the Liver building at Pier Head forms a dramatic background to this photograph, which brings together the Leyland-Albion-Scammell trio as used in the nationalised transport fleet of 1957. The old established Scottish vehicle builder Albion Motors was acquired by Leyland in 1951, and Scammell Lorries became part of the organisation toward the end of 1955.

Left: Pictured reversing onto the Ionic Ferry at Preston, is an ECO2 model 'Comet' with Briggs cab, in service with Teasdale of Carlisle, a firm of manufacturing confectioners. The ferry will dock at Larne, in County Antrim, Northern Ireland, and the vehicle soon on its way to carry out deliveries throughout Ulster.

Below: The overworked adjective 'classic' comes to mind when trying to describe the Leyland fleet of Jacob's Biscuits. The Liverpool based biscuit manufacturer had started using Leylands well before the Great War, and many readers will recall the original Matchbox toy of the 4 ton van in their livery. Typical of the 1950s fleet was this 'Hippo', with container displaying giant-sized lettering to good effect. The 'washed out' appearance of the grass has more to do with the ageing of the colour transparency, than being the result of a very dry summer!

Above: In this 1957 view of Preston docks, BRS Leyland 'Comet' artic tractors are busy marshalling trailers for the roll-on, roll-off ferry service to Larne in Northern Ireland. The vessel is the Empire Nordic, a second generation ferry operated by the Transport Ferry Service, which began operations following the end of World War 2. This was the era when large numbers of trailers used the Scammell type of automatic coupling, which was appropriate for loads up to about 12 tons.

Right: The Iberian peninsular was a good export market for Leyland products in the years following the end of World War 2, and in Spain the marque was well represented with four- and six-wheel rigids. This 1957 photograph shows an ECOS2 Comet with the O.350 engine, in service with the Hernanded Sindical de Labradores, a form of farmers' co-operative. The inclusion of the little three-wheeled truck, powered by a motorcycle engine, provides a flavour of popular local transport.

Above: Framed by overhead pipe runs, this 1958 24.04 model 'Octopus' was operated by Brown & Polson, for the delivery of their Globe Glucose, in bulk to industrial users. In order to provide a clean appearance, the lower part of the tank has been panelled, and the unloading controls located inside a cabinet at the rear. The use of spats to conceal the rear wheels completed the neat exterior, but such features do not appeal to those engaged with maintenance and wheel changing!

Left: The road haulage scene of the post-1953 Act period was originally made up of a very large number of operators, each with a few vehicles. One of these was the Davis family which had been in operation prior to nationalisation as R. Davis, the mother of the brothers who subsequently became known as Davis Brothers. In effect there were several fleets which went under the names of C. Bristow, Tozer Transport, W.D. Monger, as well as Davis Brothers. A number of 'Comet' flats were operated, one of which is seen here being sheeted up in Liverpool docks.

Right: The famous landmark of the twisted spire of Chesterfield Cathedral forms the backdrop to this shot of a Stones Brewery 20H series 'Hippo' passing through the city in December 1958. All the postwar Leyland models could be seen in brewery fleets during this period, with some being operated with drawbar trailers.

Above: This publicity photograph was probably taken after the vehicles had been in service for some time, because the first vehicle has had a great deal of cleaning and painting, while the other nine are rather less tidy! These long 'Super Hippo' artics are probably EH13 models with a wheelbase of 14ft 6in, and they are operated by Cargo Carrier (Pty) Ltd of Germiston, South Africa.

Right: A tranquil country scene, as a 14SC model 'Super Comet' of George C. Croasdale Ltd, takes on a load of round timber at Haverthwaite. The load is destined for the agricultural show at Blackpool in 1961. The listed wheelbase for a 'Super Comet' tractor of this period was 8ft 1in, but the example shown would appear to be rather longer, suggesting that a tipper chassis might have been used in order to give greater clearance behind the cab.

Below: When the LAD cab was introduced to the Leyland range in 1958 it was named the Vista-Vue. A couple of years later some of the engines were offered in an up-rated form, and these models were then known as 'Power-Plus'. This 1961 articulated Power-Plus 'Beaver' was operated by Brady & Sons of Barrow-in-Furness on behalf of British Cellophane Ltd.

Opposite page – below: Another scene at Preston docks with a BRS 'Super Comet' tractor coupled to a Van de Burgh trailer lettered for Stork Margarine, alongside the Doric Ferry. A large proportion of the vehicles using the Preston-Larne ferry service were articulated trailers, they being collected at the destination by another tractor which carried out the delivery.

Above: One of the best turned-out fleets of the 1960s was undoubtedly that of Triplex Glass Co, whose vehicles were to be seen carrying safety glass to motor industry plants around the country. This 1964 photograph shows one of the Triplex 24.O series 'Octopus' and trailer outfits leaving the factory with 28 stillages of safety glass.

Opposite page – top: A 20H series Leyland 'Hippo' tractor forms the basis for this articulated aircraft fueller, in the colour of Mobil Oil. It is pictured alongside a South African Airways Boeing 707, at Salisbury Airport, Southern Rhodesia in 1962.

Opposite page – bottom: A familiar sight around Sheffield during the 1960s, were the eightwheel bulkers operated by BRS on contract to Steel, Peech & Tozer. These two examples of the Power-Plus 'Octopus' were used to bring raw materials to the United Steel Co's works in Sheffield, where the 'World's Largest Electric Steel Plant' was located.

Right: A model 14B 'Beaver' forms the basis for this Shell BP aircraft fueller pictured alongside a Douglas DC3 of Aden Airways, at Aden Airport in 1965.

Above: Operating from their Liverpool base, the vehicles of Robertson, Buckley were regularly to be seen in the dock areas of the city. This Ergomatic cabbed 14BT/17AR model 'Beaver' with BTC tandem axle trailer, is seen whilst working under contract for Cory Associated Wharves in October 1965.

Opposite page: The cargo-handling area of London's Heathrow Airport, is the location of this evocative night shot, as the driver of a Shell BP aircraft fueller radios the control centre for instructions. Based on the 20H 'Hippo' chassis, the tanker carries pumping, metering and recording equipment for fuelling aircraft, and to provide details necessary for charging the operator.

Above: For very many years Guinness, was shipped from Dublin to Liverpool in wooden hogsheads. In more recent times bulk tanks have been used, these being originally handled by crane. Further modernisation provided fork lift trucks capable of handling the tanks, and in this 1967 view a Guinness Freightline 'Beaver' is being unloaded at St. James's Gate Brewery.

Below: A very large number of Leyland vehicles became absorbed into the fleet of British Road Services, following the passing of the Transport Act 1947. The new organisation continued to buy Leylands as the years progressed into what was the heyday of BRS in the 1950s, and to a lesser extent in the post 1953 Act period. The 'Beaver' was a natural choice for articulated outfits operating at maximum weights, and this example with York semi-trailer is seen being loaded with steel billets in 1966.

Above: This tilt-cab version of the 'Super Comet' was introduced in 1966 and was designated as 14SC, 15SC or 16SC according to application. In this photograph a 1968 model, operated by Thwaites Brewery of Blackburn, is seen at one of the company houses. Note the period TV aerials which tend to spoil the appearance of the house; items such as these were often removed before the photograph would be used for publicity purposes.

Left: The rapid growth of the market for frozen foodstuffs in the 1960s, was met by an equally rapid expansion of transport equipment to meet the demand. Although the railways had a large share of the transport of fresh fish carried in ice-filled boxes, the new trend in rapidly frozen food was met by the swifter road sector of the transport industry. Pictured against typical North Sea fishing boats, is one of the sizeable fleet of Ross Foods' LAD-cabbed 'Beaver' artics with refrigerated trailer.

Above: South America was once a good export market for British vehicles, and in postwar years the bonnetted range sold quite well. Here, a 'Super Beaver' 18EB model and trailer hauls wine for the Transports Vino Vanguardia of Argentina in 1968. The driver must have been quite tough, for the spare wheel is carried on top of the tank, alongside a number of wine flasks in raffia baskets!

Below: One of the great disappointments of the 1960s was the gas turbine truck. A handful of manufacturers carried out design and development work, both here and in the United States, but nothing ever progressed beyond the production of a few prototypes. Leyland's entry to the field was powered by a gas turbine researched and developed by Rover. Three prototypes were built for user trials, these going to Shell, Esso and Castrol.

Above: The 'Beaver' name was first used to signify the 3½ ton payload fourwheeler, when the T-engine range was introduced toward the end of 1928. After a lapse of some forty years, and with countless changes to the design, the 14BT model 'Beaver' shown here was designed to operate at 30 tons gross.

Right: A specially-adapted 'Hippo' chassis formed the basis for this heavy duty recovery vehicle, built for work in the Mersey Tunnel. Using the current tipper chassis, the rear end and bogie was suitably strengthened, a Holmes twin-boom crane added, and a locker for tools and equipment fitted. The fitment of a police-type bell behind the front towing eye is worthy of note. This vehicle stood ready for removing the heaviest vehicles from the tunnel, lighter vehicles were handled with a Land-Rover.